Make: Tools

How They Work
and How to Use Them

Charles Platt

Make: Tools
How They Work and How to Use Them
by Charles Platt

Published by Maker Media, Inc., 150 Todd Road, Suite 100, Santa Rosa, California 95407.

Maker Media books may be purchased for educational, business, or sales promotional use. Online editions are also available for most titles (oreilly.com). For more information, contact our corporate/institutional sales department: 800-998-9938 or corporate@oreilly.com.

Publisher: Roger Stewart
Copy editor and proofreader: Brian White
Interior design: Erico Platt
Cover design: Julie Cohen
Cover photographs, interior photographs, and diagrams: Charles Platt

October 2016: First Edition

Revision History for the First Edition

2016-10-01 First Release
2017-10-13 Second Release
2023-09-08 Third Release

See oreilly.com/catalog/errata.csp?isbn=9781680452532 for release details.

ISBN: 978-1-680-45253-2

Dedication

For Mark Frauenfelder and Kevin Kelly, who planted the idea, and Jeremy Frank, who helped so much with the implementation.

Acknowledgments

I appreciate the feedback from Mark Jones, Marshall Magee, Russ Sprouse, and Gary White.

My editor, Roger Stewart, was endlessly helpful in making this book possible.

Contents

Introduction

Some skills are so basic, people take them for granted. Hammering a nail into a piece of wood, for instance. What could be simpler than that?

Actually, it's not so simple. There are more than 20 kinds of hammer. Can you recognize the difference between a **claw hammer** and a **cross-peen hammer**? And does it matter?

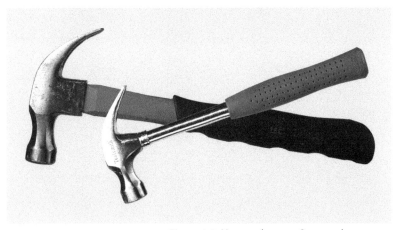

Figure I-1. You can have an 8-ounce hammer, a 16-ounce hammer, or something larger, if you prefer.

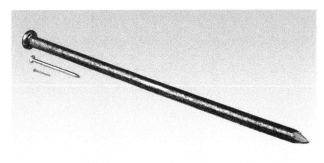

Figure I-2. Yes, you really can buy a 10-inch nail (although it is properly known as a "spike"). A 1/2" nail can be sold as a "brad." The one in between is just a nail.

The head of a typical hammer weighs 16 ounces, but you can buy an 8-ounce hammer if you prefer (see Figure I-1). Or how about a 24-ounce hammer? Which type is easier to use? And would a fiberglass handle be better than a wooden handle? How about a steel handle? And when you hold the handle, should you grip it near the head of the hammer or near the end of the handle?

Now consider your choice of a nail. There are **common nails**, **finishing nails**, **galvanized nails**, **ring-shank nails**, **coated nails**, **decorative nails**, and many more. In the United States, common nails are often sold by weight, and their size may be described in pennies. If you buy a pound of sixpenny nails, would you like to know how big they are and how many will be in the box?

What's the largest nail you can buy? (It may be larger than you think. See Figure I-2.)

Suppose you want to cut some wood. This raises many more questions. If you use a **handsaw**, how do you start the cut without the saw jumping around or sliding sideways? Is a Japanese-style **pull-saw** easier to use than a Western-style saw that cuts when you push it? Will you be more likely to hurt yourself with a pull-saw?

How can you prevent the underside of a piece of wood from splintering when the saw emerges through it? How do you make a precisely vertical cut? Power saws are easier to use in many ways than hand saws, but in that case, why do handsaws still exist? If you use a power saw, should it be battery-powered or plug-in?

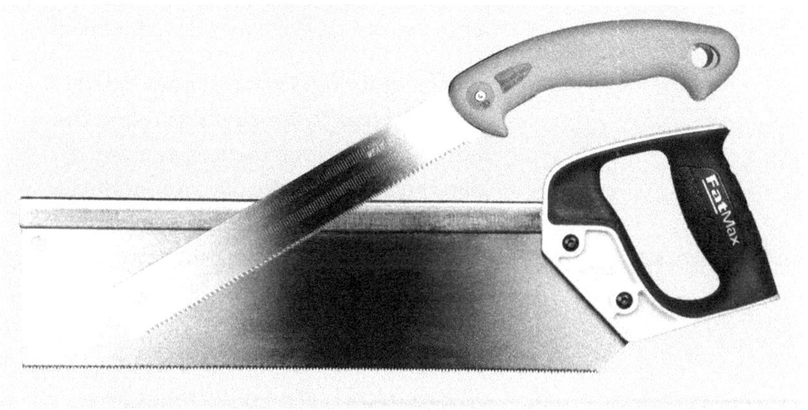

What are the relative advantages of a **reciprocating saw**, a **circular saw**, a **band saw**, a **scroll saw**, a **jigsaw**, and a **table saw**? How many teeth per inch should the blade have? Do you need a different type of blade to cut plastic?

Figure I-3. An 8½" Japanese pull-saw, or a 14" tenon saw—which do you prefer?

Really, there are so many questions about tools, you need a whole book to answer them.

This is that book.

Are Tools Still Useful?

In a world where **3D printing** can create anything from a gearbox to a house, you may wonder if anyone needs hand tools in a workshop anymore. It's true that 3D printers are wonderful, but they have limitations. For a start, none of them can work with wood, and very few can use metal. They are ideal for prototyping, but a component fabricated with a 3D printer will almost always have to be mated with some other part, or may need to be installed in an enclosure or adapted in some way. For these purposes, workshop skills are still useful.

Moreover, if you want to make something that is beautiful instead of merely functional, you will need to work with your hands.

Who Can Use This Book

Anyone who is willing to follow simple instructions can use this book, so long as you have basic manual ability and reasonably good eyesight. Age and gender are not relevant.

You can be a complete beginner, but if you do already have some basic skills, I think you'll still find a lot of value here. I will be bringing together numerous facts, tips, and tricks that have taken me a lifetime to learn. You may not know all of them.

This book will also serve as a reference source. If you want to remind yourself of the difference between a tenon saw and a panel saw, or if you are trying to remember if poplar is a harder wood than birch, you can look up the answers here.

Learning by Discovery

Several years ago I wrote a book titled *Make: Electronics,* which encouraged the reader to build little devices as a way of learning about electronic components. I began with the simple task of switching on an LED, and worked up to ambitious projects such as a burglar-alarm system. Because the reader discovered basic principles by using this hands-on approach, I called it "learning by discovery." I think it is by far the best way to learn.

Figure I-4. Tool usage is not age-specific, gender-specific, or anything-else-specific.

This book uses the same plan. You will begin with very simple projects, such as a sliding-block puzzle. You may end up making a jewelry box, a whistle, or a geodesic lamp shade. (Maybe you don't think you need a geodesic lamp shade—but when you find out what it is, you may like it.) Along the way you will learn about different types of tools, how to use them, and how you can make mistakes that mess things up.

Mistakes are an important part of the learning process. Cutting wood to the wrong length, splitting it by banging a nail into it (as in Figure I-6), or mashing the head of a screw—everyone does these things, and no one should be embarrassed about them. In fact, you need to make mistakes, so that you can learn how to avoid them.

Figure I-5. Anyone can learn to hammer nails. Even a clumsy weird person.

Figure I-6. Everyone does this at least once. Some of us, more than once.

You also need to see what happens if you don't follow instructions. For instance, most books will tell you that after you apply wood glue, you should clamp the parts for 24 hours. But what if you don't clamp them? Or what if you only clamp them for one hour? The best way to find out is by trying it.

What This Book Doesn't Do

If you want to build large, complicated, elegant pieces of furniture in a basement workshop equipped with $10,000 worth of power tools—this book is not for you. Many books and magazines on woodworking deal with that kind of advanced scenario.

If you want to lay ceramic floor tiles, install drywall, or do some plumbing work, I will not be dealing with those tasks. You need a DIY home improvement book for that. Many of them are available.

What You Will Need

Because the projects in this book are small in scale, you don't need a workshop to build them. You don't even need a workbench, if you have a sturdy table. That's how I used to build things myself, when I lived in a tiny apartment in New York City. My work area (around the table) was about 72" x 72". I managed, and you can, too. Of course, if you do have a workbench, that will be more convenient.

With the exception of an electric drill and electric screwdriver, the projects use hand tools because they're cheaper and less dangerous than power tools. If you want more capabilities, Chapter 20 includes a list of additional tools that you may want to own.

I have done my best to minimize the expenses for materials and accessories.

Will You Hurt Yourself?

Every crafts hobby entails a little bit of risk. You can even get hurt while sewing, as I know through my own experience. Because I was careless and impatient, I managed to cut off a small piece of the tip of my finger with a pair of scissors.

The key words, here, are "careless and impatient." When you're impatient (as I tend to be), it is a risk factor.

Pause frequently. Think ahead. When you find yourself making silly mistakes, take a break. If you can work in this way, you should be able to avoid cuts or bruises.

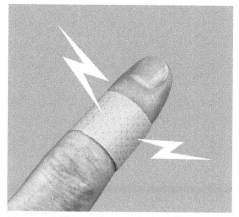

Figure I-7 This can be avoided. Really.

What Will You Build?

This book contains more than 20 projects that I describe in enough detail to enable you to build them yourself. They are sequenced so that additional skills, tools, and accessories are involved as you move through the book. More projects are included in summary form.

You'll find some traditional items, such a jewelry box and a child's toy monster truck, but I also tried to come up with unusual things that you may have never imagined making.

Along the way you'll find that I return several times to the topic of geometry. This is because the process of making things is all about shapes and sizes, and in a task such as cutting mitered corners of the type used in a picture frame, you can't get away from the concept of angles. You can skip my geometrical digressions if they don't interest you, but a little basic geometry will help you to design things of your own.

Why Build It, When You Can Buy It?

Some of the items in this book can be bought from stores for very little money. Why waste time making your own?

I believe you can derive unique satisfaction from making things. Being able to point to an object and say, "I made that," is a special feeling. The object becomes an extension of yourself that will endure for decades, and it can be a memorable gift, because everyone likes to receive something that is hand made.

Additionally, I believe that you can benefit from feeling capable rather than helpless when dealing with the modern world. If you can put up some book shelves or hang a picture without depending on someone for assistance, that's a good feeling. You can also apply your skills to repair things when they break, or modify gadgets so that they serve a purpose that the manufacturer never imagined.

Lastly, as everyone knows, if you do a job yourself, you are more likely to get it done the way you want it.

My Own Skill Set

You may be wondering why I feel qualified to write this book. Long ago, I graduated with a score of 99 percent from an advanced woodworking class, but I don't regard that as a primary qualification. I learned more from my father, who was an automotive engineer. He taught me to use tools at an early age, and I've been designing and building things ever since, ranging from small toys to outbuildings.

I taught a college course titled "How Things Work and How to Fix Them." I have customized vans, fabricated electronic devices, and renovated a kitchen. More recently, I spent five years designing and building prototypes for a California laboratory.

As a writer, my name is on more than 40 books, on topics including computer programming and electronics hardware. I also wrote a book about decorating your own T-shirts. Currently I'm a contributing editor for *Make* magazine, for which I have written more than 50 features. I served as a guest editor for a web site named *Cool Tools*, and compiled "Tool Tips" for the same site.

To fill the gaps in my own knowledge, I have a very valuable advisor, Jeremy Frank, who has spent much of his life working as a professional machinist. I rely on Jeremy as my consultant, and for clever little tips that he picked up over the years. Between the two of us, I think we can cover tool usage fairly thoroughly.

The Goal

Much of my knowledge has been gained the hard way: by trial and error. I have also collected bits and pieces of information by reading books, catalogs, and web sites, and by visiting stores, and asking people.

You could learn about tools in the same way that I did, on a piecemeal basis. But wouldn't it be easier if all the information was gathered for you in one place, and organized so that it will be easy to learn? That is the goal of this book.

Errors and Questions

If you find an error in this book, please report it so that we can fix it in future printings. The system for this is very well established. Just go to this URL:

http://www.oreilly.com/catalog/errata.csp?isbn=0636920031154

If you have questions for the author, the situation is a little different. I don't have time to answer every question about tool usage, but if you feel that something in the book isn't clearly explained, I encourage you to let me know. The email address for this purpose is platt.tools@gmail.com. I personally read all messages sent to that address. Sometimes I can reply immediately, while other times it may take me 10 days. Please be patient!

If you want news and updates about my other books (past and future), and some links that may be useful, please visit www.plattelectronics.com.

Complaints?

You have a lot of power as a reader. One negative review on Amazon can outweigh a dozen positive ones. Therefore, if you have a complaint, please let me know, in case there has been a misunderstanding, or in case I can resolve the issue in some way. Just give me a chance to make it right before you think of taking it public. Thanks.

Safari® Books Online

Safari Books Online is an on-demand digital library that delivers expert content in both book and video form from the world's leading authors in technology and business. Technology professionals, software developers, web designers, and business and creative professionals use Safari Books Online as their primary resource for research, problem solving, learning, and certification training.

Safari Books Online offers a range of plans and pricing for enterprise, government, education, and individuals.

Members have access to thousands of books, training videos, and prepublication manuscripts in one fully searchable database from publishers like O'Reilly Media, Prentice Hall Professional, Addison-Wesley Professional, Microsoft Press, Sams, Que, Peachpit Press, Focal Press, Cisco Press, John Wiley & Sons, Syngress, Morgan Kaufmann, IBM Redbooks, Packt, Adobe Press, FT Press, Apress, Manning, New Riders, McGraw-Hill, Jones & Bartlett, Course Technology, and hundreds more. For more information about Safari Books Online, please visit us online.

How to Contact Us

Please address comments and questions concerning this book to the publisher:

Maker Media, Inc.
1160 Battery Street East, Suite 125
San Francisco, CA 94111
877-306-6253 (in the United States or Canada)
707-639-1355 (international or local)

Maker Media unites, inspires, informs, and entertains a growing community of resourceful people who undertake amazing projects in their backyards, basements, and garages. Maker Media celebrates your right to tweak, hack, and bend any Technology to your will. The Maker Media audience continues to be a growing culture and community that believes in bettering ourselves, our environment, our educational system—our entire world. This is much more than an audience, it's a worldwide movement that Maker Media is leading. We call it the Maker Movement.

For more information about **Maker Media**, visit us online:

- Make: and Makezine.com: makezine.com
- Maker Faire: makerfaire.com
- Maker Shed: makershed.com

To comment or ask technical questions about this book, send email to bookquestions@oreilly.com.

Chapter 1
A Perplexing Puzzler

I'm going to start with something that I believe absolutely everyone can make. All you have to do is cut a length of wood into pieces, glue two of them together, and round their edges with sandpaper. Add a piece of cardboard, and the job is done.

The end product will be a little game named Dad's Puzzler, which was invented in England more than 100 years ago. In fact, at that time, it became a national obsession. If you'd like to know why everyone became so excited about it, building a copy for yourself will only take an hour or so. You'll find that it's fun to play with—and very challenging.

The game consists of a tray containing nine wooden blocks, as shown in Figure 1-1. The idea is to slide the blocks around in such a way that the big square block ends up in the bottom-left corner.

NEW TOPICS IN THIS CHAPTER

- Making measurements
- Sawing with a miter box
- Smoothing edges with sandpaper
- Gluing and clamping
- Finishing with polyurethane

See next page for a materials list.

That sounds easy, but you'll find that it takes more than 40 moves. Naturally you're not allowed to rotate the pieces or lift them out of the tray.

The tools and materials that you will need are listed on the next page. You may not know what some of these things are, yet, or how to use them. But if you build this project, you will quickly find out.

Figure 1-1. The concept of Dad's Puzzler is simple enough. Solving it is another matter.

YOU WILL NEED

- Tenon saw with hardened teeth
- Miter box
- Trigger clamps, quantity 2
- Ruler, 18", stainless steel, cork back, calibrated in inches and millimeters
- Speed square, 7"
- Rubber sanding block
- Work gloves
- Dust mask (optional)
- Safety glasses (optional)
- Square dowel, ¾" x ¾", length 36"
- Plywood, ¾" or less, as a work surface, size 24" x 24" or larger (Masonite is acceptable)
- Carpenter's glue, 8 ozt
- Sandpaper, 80 grit, 9" x 11" sheets, quantity 3
- Polyurethane, 1 pint
- Cardboard, any thickness, size 9" x 9"
- Disposable gloves for applying polyurethane
- Paint brush if preferred

Many of these items will be useful in future projects.

I assume you already have a pencil, paper, eraser, and paper towels.

Throughout this book, the " mark is used to mean "inches." Units of feet are not used.

Check the Buying Guide on page 248 for information about buying these items.

Buying a Square Dowel

At your local big-box hardware store, I'm betting you will find some **square dowels**. People tend to think that a dowel is always round, like the rod in a clothes closet. But it can also be square, when viewed from the end, as shown in Figure 1-2.

Allowing for some mistakes and wastage, and including a frame around the blocks, you will need 36" of dowel to build Dad's Puzzler. When you buy the wood, check it carefully to make sure that it is straight and free from defects.

What if your local store doesn't have wood of this type? You can buy it online. Please see page 248 where the Buying Guide will provide advice.

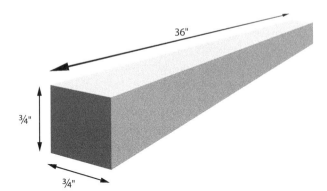

Figure 1-2. For this project, a square dowel must be used. The type of wood is not important.

Establishing a Work Surface

A workbench is nice, but not necessary in this book. You can use a kitchen table, so long as it is strong and rigid. To protect its surface, I suggest a piece of **Masonite**, or **plywood** ranging from ¼" to ¾" thick. It should be at least 24" x 24". Some stores sell precut pieces 24" x 48", which are ideal. If you can only find a full-size sheet measuring 48" x 96", the store should be willing to cut it into pieces for you.

For information about composite materials such as plywood and Masonite, see the Composites Fact Sheet on page 78.

Choosing a Saw

I'm going to ask you to use a **handsaw**, not a **power saw**.
There are four reasons:
1. The hand-eye coordination that you will learn with a handsaw will be transferable to other tools, and is valuable.
2. For small, detail work, you often have to use a hand saw.
3. Hand saws are generally cheaper than power saws.
4. Hand saws tend to be safer than power saws.

If you want to use power saws later, that's up to you. You'll find a discussion of the types that are available on page 245. But I don't think it's a good idea to use them without an experienced person guiding you initially.

Because handsaws are such basic tools, many variants exist. See the Handsaw Fact Sheet on page 16 for details. For our purposes, you need:

Tenon saw with **hardened teeth**, 14" or shorter, 12 or 13 teeth per inch. See Figure 1-3.

For faster cutting of larger pieces of wood, you may benefit from a second saw:

Utility saw with **hardened teeth**, 10" approximately, 10 or 11 teeth per inch. See Figure 1-4.

The lengths are of the blade, not of the complete saw.

A **tenon saw** is also known as a **miter saw**, because it can be used to make joints called **miter joints** (which I'll deal with later). Some people may call it a **back saw**, because it has a stiffening bar along its back. This saw is good for making precise cuts in small pieces of hardwood.

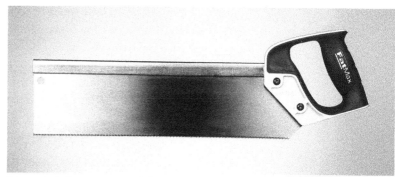

Figure 1-3. A tenon saw with hardened teeth.

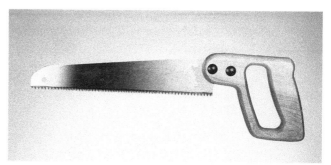

Figure 1-4. A utility saw.

A **utility saw** is sometimes called a **toolbox saw**, because it's small enough to fit in a toolbox. It is not as precise as a tenon saw, and does not make such clean cuts, but it does cut faster.

The utility saw is optional. You can make all the projects in this book with a tenon saw. However, it must have hardened teeth, because their shape and sharpness will greatly reduce your effort. The only disadvantages of hardened teeth are that they are more brittle, and resharpening the saw with a metal file is difficult—but not many people try to do that, anyway.

For reasons that I do not understand, most tenon saws do not have hardened teeth. You may have to search online to find one. At the time of writing, the Stanley FatMax 17-202 is a great bargain, and is exactly what you need. See page 251.

Stanley also makes a utility saw, model 20-221.

Handsaw Safety

If you wrap your hand around the blade of a saw and grip it gently, you are unlikely to hurt yourself, so long as you don't squeeze too hard. This is because the force of your grip is spread across dozens of saw teeth, and the teeth are not moving across your skin.

Don't let this give you a false sense of security. When the saw is moving, those teeth rip into wood with surprising power, and can cut your fingers just as easily.

Here are the safety rules:

■ Keep your free hand at least 4" from the saw, if you are sawing without a miter box. A clumsy saw stroke can cause the blade to jump sideways and cut your finger.

■ Wear **work gloves**. They can interfere with small, precise movements, but will provide valuable protection (from handsaws, but not sufficient to protect you from power saws). Any hardware store sells a variety of work gloves.

■ If you find your eyes getting sore and gritty, protect them from sawdust with **safety glasses**. If you find yourself coughing, protect your lungs with a **dust mask**.

■ Never place any part of your hand under the blade while cutting. See Figure 1-5.

■ Never use a saw after drinking or while under the influence of drugs.

■ Children should be supervised.

■ When you finish using a saw, put it in a drawer or box, or hang it up securely, using the hole in the blade

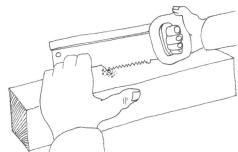

Figure 1-5. Does anyone see anything wrong with this picture?

that is provided for this purpose. If you leave a saw lying on a bench, you can knock it accidentally onto the floor, where it can land teeth-down on your foot. This is bad for the saw and bad for you, if you're wearing sneakers or (worse) sandals. Work shoes are recommended in a workshop. Sandals should never be worn. Bare feet are out of the question.

Hand saws can be safe, so long as you use them carefully.

What Is a Miter Box?

This device helps you to make straight, correctly angled cuts. You can think of it as being like training wheels on a bicycle. In Chapter 3, you'll be making your cuts freehand, but it's more difficult. A yellow plastic miter box is shown in Figure 1-6.

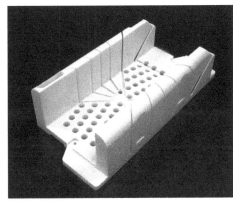

Figure 1-6. A miter box will help you to make precisely angled cuts.

At your local retailer, you may see a miter box and a tenon saw sold in one package. However, in the combination that I tested, the saw was of poor quality, did not have hardened teeth, required a lot of muscle power, and was not very useful. You may spend a few extra dollars to buy a saw separately, but I think it is worthwhile.

Clamping Your Work

Many people will tell you that you can saw wood by placing it on a saw horse and holding it there with your foot. I've worked this way myself, but accuracy is difficult to achieve when you are standing on one leg, anchoring the wood with your other leg, and working at arm's length.

For the projects in this book, you will need to make precise cuts. The best way to achieve this is to clamp your work to a solid bench or table.

A **trigger clamp** is the easiest to use, as shown in Figure 1-7. It is also known as a **bar clamp**. The small metal lever is the trigger, which releases the jaw of the clamp, allowing it to slide up or down the bar. Let go of the trigger, and the large black plastic lever closes the jaws of the clamp when you squeeze it repeatedly.

A **screw clamp** is more powerful, but using it is more time-consuming.

Two trigger clamps are all you need for the projects in this book. If you are using a miter box, Figure 1-8 shows how to clamp it to stop it from jumping around. Anytime you are not using a miter box, you can apply a clamp directly to the wood that you're working on.

Cutting a Usable Length

Long pieces of wood are difficult to control precisely. If your piece of square dowel is longer than 36", you will have to reduce it to 36" or less.

Place the dowel in the miter box as shown in Figure 1-9.

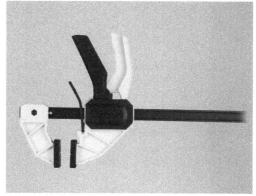
Figure 1-7. A trigger clamp is essential to stabilize your work.

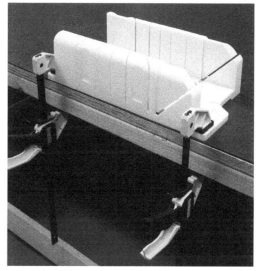
Figure 1-8. Your miter box should have provision for you to clamp it in position.

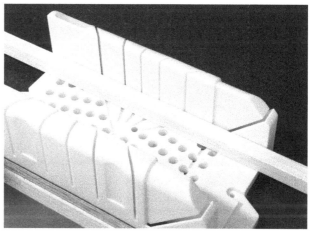
Figure 1-9. Placing a square dowel in the miter box.

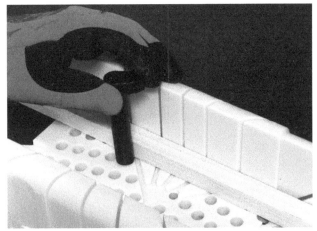

Figure 1-10. Some miter boxes provide cams that you turn to hold the wood in position.

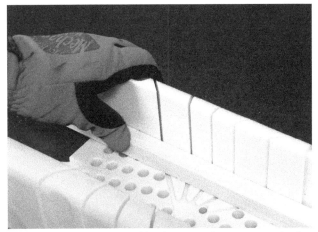

Figure 1-11. Hold the wood firmly against the opposite side of the miter box.

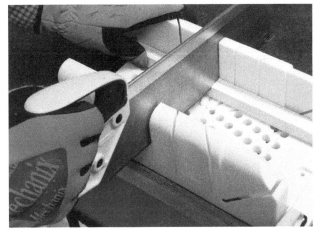

Figure 1-12. Sawing wood with a miter box.

The box that I have is equipped with a couple of pegs, described as **cams**. You insert a cam in a hole close to the wood and turn it to lock the wood in place, as shown in Figure 1-10.

If your miter box doesn't have this feature, you can hold the wood in the miter box with your left hand, as shown in Figure 1-11. (If you are left-handed, hold it with your right hand.). Be careful not to put your hand too close to the saw. I do recommend that you wear work gloves while using a saw.

Beginning a cut can be difficult, because the saw tends to dig in. Draw the saw toward you a few times, to create a shallow groove in the wood. Now when you push the saw, it should cut more easily. If you still have trouble, drag the saw toward you a few more times. Figure 1-12 shows a cut in progress.

Don't press too hard when you are cutting the wood. You shouldn't be fighting a battle. The saw should do most of the work for you.

Trimming the End

Assuming you now have a piece of square dowel that is 36" or shorter, you need to trim a thin slice from the end of it. This is a little routine that you should follow, anytime you begin using a piece of wood that is fresh from the store. Always remember:

▪ The end of the wood that you buy may not be trimmed accurately. It could have a slight angle.

You don't need to measure the amount that you will remove in this cut. About ¼" will do. Follow the same procedure that I described in the section above.

If you don't succeed in making a nice clean, square end on the dowel on your first attempt, move it along a little and try again.

When the end of the wood looks good, you're ready to cut a measured section to use in the block puzzle. So, you have to measure it.

Making Measurements

Your most essential measuring tool is a **stainless-steel ruler**. Why not use a cheaper one, made of plastic? Because it will tend to get chipped, scuffed, or broken in a workshop environment. Stainless steel also allows you to make finer measurements, usually down to 1/32".

Some stainless-steel rulers have an extra margin at the ends, like the upper one in Figure 1-13. Others begin measuring from the end of the steel, like the lower one in Figure 1-13. I prefer not to have a margin, as it often gets in the way—for example, if you stand a ruler inside a box to measure its depth. If you shop for a ruler without a margin, make sure that the scale begins precisely at the end of the metal. Sometimes during the manufacturing process, the measurement scale is not applied accurately.

Your ruler must be 18" long and should have a cork back to stop it from sliding around. This is more important than it sounds. The back is shown in Figure 1-14.

The ruler must have a scale in millimeters as well as inches. In the United States, when cutting wood, we almost always use inches divided in half, and half again, and half again. This means that dimensions are usually expressed in multiples of 1/2", 1/4", 1/8", 1/16", 1/32", and 1/64". Figure 1-15 is my attempt to make this clear.

When you are using fractions, if you double the number at the top and also double the number at the bottom, the value stays the same. This means that 2/4" would be the same as 1/2", and 6/32" would be the same as 3/16".

Knowing this, you can figure out if a value such as 5/32" is bigger or smaller than 1/8". Using the doubling rule, 1/8" is the same as 2/16", which is the same as 4/32". So, 5/32" is bigger than 1/8". This kind of thing is important when you are using a drill bit to make a hole, and you want the hole to be just a fraction bigger.

I included hundredths of an inch in the diagram, because sometimes that system is used (for example, when measuring the thickness of metal). And I included millimeters, just for the sake of completeness.

Note that there are precisely 25.4mm in an inch.

Figure 1-13. A stainless steel ruler may have an extra margin at each end, shown in the upper sample here. A zero margin, shown in the lower sample, is preferred.

Figure 1-14. The cork back of a stainless-steel ruler.

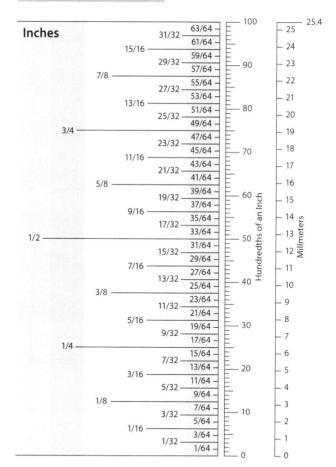

Figure 1-15. Units of measurement.

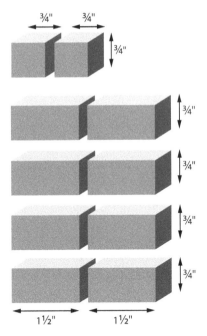

Looking back at Figure 1-1, you'll see that the puzzle consists of two small square blocks, six rectangular blocks (each of which is twice as long as a small square), and one big square block. How are you going to make them?

Each of the small squares can be a piece of ¾" dowel that is ¾" in length. Each of the rectangles has to be twice as long, which is 1½" in length. But what about the big square, which is 1½" x 1½"?

No problem. You will cut two additional pieces measuring 1½" long, and glue them together, side by side, to make a square.

If you add it up, you will actually need a total of eight pieces measuring 1½" each, and two pieces that are ¾". This is illustrated in Figure 1-16.

But don't start measuring yet! When a saw makes a cut, it always removes some of the wood. If you mark all the distances on your wood at one time, you won't be allowing for the widths of the cuts, and the pieces will turn out to be a little shorter than they should be. The procedure is to mark one measurement, then cut outside of that line, then make another measurement, and so on. The method is explained in the next two pages.

Figure 1-16. To create the puzzle pieces, you need these 10 sections of ¾" dowel.

Creating a 90-Degree Angle

First you need to know how angles are measured. Figure 1-17 shows some examples. An angle of 90 degrees is also known as a **right angle**. When the end of a piece of wood is at 90 degrees to its length, we often say that the end is **square**.

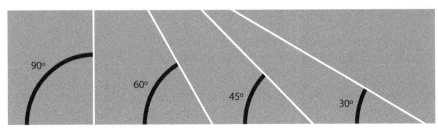

Figure 1-17. Some commonly used angles.

The miter box automatically creates cuts at 90 degrees, but sooner or later you'll have to make a cut without its help—for example, if a piece of wood is too big to fit in the box. You also need to be able to check that your cut is accurate, even if you are using the box. Therefore, you need to know how to draw a guideline at 90 degrees.

To draw the line, you use a pencil. It should be HB or harder, because a soft pencil doesn't last long on the rough surface of wood.

Wouldn't the line created by a pen or a fine-point marker be easier to see? Well, yes, but the ink will tend to sink into the wood, and will be difficult to remove later. You can get rid of pencil marks easily with a soft eraser, or by sandpapering them away.

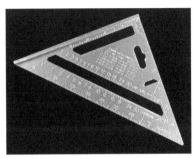

Figure 1-18. A 7" speed square. Some variants are larger, but this is adequate for our purposes.

Two tools are commonly used to make 90-degree marks. One is shown in Figure 1-18. Even though it is triangular, it is known as a **speed square**, because it helps you to make things square—that is, at 90 degrees.

Figure 1-19 shows one way to use it. Make sure the edge of the thick side of the tool is tight against the edge of the wood, and then use your ruler to establish the distance between the speed square and the end of the wood. (There are other ways to do this. If your ruler has a margin at the end, use the ruler on its own, make a mark, then use the speed square to guide the pencil when you extend the mark across the wood.)

While keeping a firm grip on the speed square with your left hand, remove the ruler with your right hand and draw a line along the edge of the speed square, as shown in Figure 1-20. Make sure there is no gap between the line and the speed square.

Another tool that you can use for this purpose is a **try-square**, shown in Figure 1-21.

You can use your steel ruler in combination with a try-square, in the same way that you used it with a speed square. Just align the edge of the handle of the try-square with the edge of your wood, as shown in Figure 1-22. However, some try-squares are not accurately manufactured. When you buy one, check its 90-degree angle against the corner of a sheet of paper or another reliable source.

Personally I prefer a speed square because I tend to be clumsy, and a speed square is less likely to be damaged if I drop it on a concrete floor. You'll see photographs of me using a speed square throughout the book. But you can use a try-square if you prefer. You'll probably find that they cost about the same.

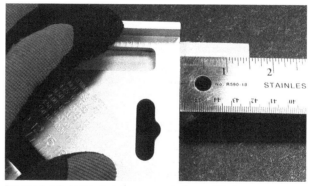

Figure 1-19. Measuring the amount of wood that you want to cut.

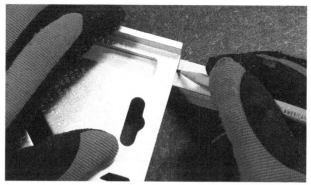

Figure 1-20. Drawing the line that will guide your saw cut.

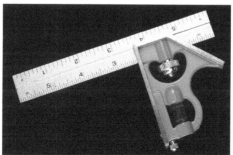

Figure 1-21. A try-square.

Figure 1-22. Drawing a 90-degree line with a try-square.

It's very important to bear in mind that pencil lines and saw cuts have some thickness. Figure 1-23 shows what I mean. The edge of your speed square (or try-square) is at the distance we want from the end of the wood. Your pencil line is inside that edge. When you remove the speed square, your saw should cut outside of the pencil line. Otherwise, the width of the saw will remove some wood from the piece that you are creating, and it won't be the right length.

Figure 1-23.
The importance of cutting outside your pencil line.

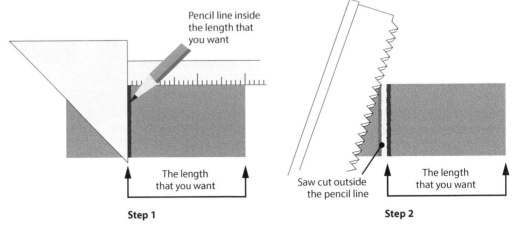

Pencil line inside the length that you want

The length that you want

Step 1

Saw cut outside the pencil line

The length that you want

Step 2

Some points to remember:

■ The point of your pencil must be sharpened frequently. Thick lines are no help in doing accurate work. You can use a mechanical pencil, if you prefer. This contains a long lead that you advance by pressing a button. You'll find it in a stationery store.

■ When you are doing precise work, it's a good idea to cut pieces a fraction larger than they should be. Then you can sand them down to the exact size.

Shaping

After you cut the wood, you'll need **sandpaper** to smooth the edges. It's available in standard-size sheets measuring 9" by 11", and in various compositions such as aluminum oxide, any of which are satisfactory for this book. Alternatively you can buy a **sanding sponge** that can be rinsed and reused. I like sanding sponges, especially when sanding curved objects where the sponge will flex to conform to the curve.

Figure 1-24. Closeup of a piece of sandpaper.

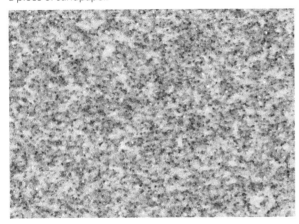

The surface of a sheet of sandpaper is shown, closeup, in Figure 1-24. A sanding sponge is shown in Figure 1-25.

The roughness of sandpaper is known as its **grit**, with 40 grit being very coarse and 600 grit being extremely smooth. You use a higher grit number to get a very delicate finish, and a lower grit number if you want to get quicker results, especially with the kind of hardwood that is often used in dowels. For this project, I suggest 80 grit.

To use sandpaper, you really need a **rubber sanding block**, which is a chunk of rubber curved on one side and flat on the other. A standard sheet of sandpaper can be divided into four equal strips, one strip being just the right size to wrap around the flat surface of the block. Don't try to cut the sandpaper—it will blunt any knife or scissors. Crease it, then reverse the crease, then tear it on the crease. You may find it helps to tear it along the edge of a table or along the edge of a ruler (but only if the ruler is placed against the back of the sandpaper, so that the ruler does not get scratched by the grit).

The end of the sandpaper goes into a groove in the block, where it is retained by three sharp pins. You can see one end of the sandpaper inserted in the block in Figure 1-26, and the pins are visible in Figure 1-27. Dig your fingers into the groove in the block, and you can pull the edge upward just enough to slip the sandpaper in. When you squeeze the block, the pins grab the sandpaper.

Now you can insert the other end of the paper in the opposite end of the block, and the result should look like Figure 1-28.

Do you really need a sanding block? Can't you just hold the sandpaper in your hand, and rub it over the wood? Sure, you can do that—but the paper, being paper, may eventually split and fall apart. The sanding block preserves it, and also gives you a relatively flat working surface.

If you were making a large object, such as a piece of furniture, you would rub the sanding block over it. For a small job like this, may be easier to rub the pieces over the block.

Experiment to find which method works best for you.

After you sand each piece, check its size, and then measure and cut another piece. Does this sound a bit repetitive? Well, yes, a lot of workshop tasks involve repetition. But if you immerse yourself in the work, you may find that it's very pleasurable, as you transform raw materials into something that is useful, pleasing to look at, and reflects the care that you apply.

Figure 1-25. A sanding sponge.

Figure 1-26. After tearing a strip of sandpaper 2¾" wide, insert one end in the sanding block, pulling up the rubber so that the sandpaper can get into the groove.

Figure 1-27. Pins inside the sanding block retain the sandpaper.

Figure 1-28. The sanding block ready for action.

A Glue Test

Lastly, you will need glue. **Carpenter's glue** is like Elmer's glue, except that it is yellow instead of white. Several brands are available. A sample is shown in Figure 1-29.

While it's wet, it dissolves in water, so you can easily remove it from your fingers. But after it sets, it can be surprisingly strong.

The two basic rules when gluing wood are:

- Remove all sawdust. Make sure the surfaces are clean and dry.
- Don't use too much glue!

Figure 1-29. A bottle of carpenter's glue. The brand that you use is not important.

The glue has to penetrate the wood slightly, and dust will interfere with this (grease is even worse). First get rid of the dust in your work area. Vacuum it away, and then put some clean scrap paper on your work area. Now wipe the pieces to be glued with a damp paper towel or rag that will pick up any residual sawdust. Leave the pieces to dry for an hour.

Maybe you should conduct an experiment before you glue the blocks that you cut. Try gluing two lengths of wood without a clamp, as shown in Figure 1-30. Just spread a thin film of glue on each piece where it will touch the other. You can use your fingers to do this, or a piece of paper towel.

The glue doesn't set very fast, so you can take your time. Align the pieces, stand one on top of the other, and leave them for a couple of hours. I think you'll find that when you come back, you can easily snap them apart.

Use sandpaper to remove the dry residues of glue, then clean the surfaces again, apply fresh glue, and hold the pieces with a clamp, as shown in Figure 1-31. Now when you come back after a couple of hours and release the clamp, you should find that the joint is much stronger.

Will the joint be even stronger if you clamp it for 24 hours? I leave that to you to find out.

Now that you've performed your test, it's time for the final joint. Align two of your blocks that are 1½" long, and apply the clamp as shown in

Figure 1-30. Testing glue strength without using a clamp.

Figure 1-31. Testing glue strength with a clamp.

Figure 1-32. Tighten it as much as you can. Once again remove any surplus of glue, and leave it to dry.

Note that glue sets faster when the temperature is higher. Below around 50 degrees Fahrenheit, it doesn't work very well.

Finishing With Polyurethane

Sand the surfaces of the pieces, and round their edges by turning them as you rub them over the sandpaper. Clean dust off them with a wet rag or paper towel, and allow them to dry.

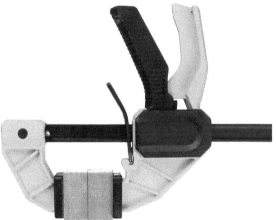

Figure 1-32. Clamping two sections of dowel to make the large square piece in the puzzle.

Now I think you should coat them with **polyurethane**. This is because bare wood tends to pick up dirt, and is difficult to clean. Polyurethane protects it, and is available in various forms: **water-based** or **oil-based**, **gloss finish**, **semi-gloss**, or **matte finish**, and **clear** or **tinted**.

Personally I prefer the oil-based type, because I think the finish is tougher, and it does not require so many coats. On the other hand, it smells bad, and you'll have difficulty cleaning your brush. But on the other, other hand, you can use a disposable brush that costs 50 cents or less. The choice is yours.

For this project, I think a clear semi-gloss polyurethane is appropriate. A 1-pint can will be enough for all the projects in the book.

The basic rules for applying polyurethane are very simple:
- Don't shake the can! Bubbles will form in the liquid, and will spoil the finish. Stir gently.
- Two thin coats are always better than one thick coat, which will tend to run.
- Wear gloves. You may think you are going to be so careful, you cannot possibly get polyurethane on your fingers. Well, good luck with that! I think it's easier to use disposable gloves. These should not be latex gloves, because contact with latex can eventually give you an allergy.
- While the coat of polyurethane is drying, don't leave it anywhere that insects can land on it.
- Keep your work away from pets or children.
- Polyurethane dries faster in a higher ambient temperature, with good ventilation. The ventilation is also important to minimize inhalation of the fumes. If you use a fan, however, be careful that it doesn't stir up dust.

Should you use a **bristle brush**, or a **foam brush**? This is a matter of taste. A cheap bristle brush may shed bristles that you have to remove carefully from your work before it dries. But a foam brush is not robust, and can shed foam particles from its wedge-shaped edge. Personally I prefer a bristle brush, because the bristles are more easily removed than the foam particles. But some people prefer to apply polyurethane with a rag. If you've never done this, it's worth a try.

Figure 1-33.
Frame dimensions
for Dad's Puzzler.

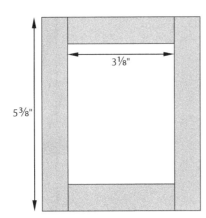

3⅛"

5⅜"

If you use a brush and wrap it tightly in a food storage bag after applying one coat, the brush should still be wet enough when you unwrap it to apply another coat.

Framing the Puzzle

If you look back at Figure 1-1, you'll notice that our job is not quite done. We still need to make a frame to contain the puzzle.

One way to do this is to cut a narrow strip of cardboard, wrap it around the pieces to form an enclosure, and tape it in place. Then add a cardboard base and secure it with more tape.

But maybe you would prefer a better-looking frame. You can use more of your square dowel for this purpose, with measurements shown in Figure 1-33. I have added a little extra to the frame measurements, so that the blocks have room to move around.

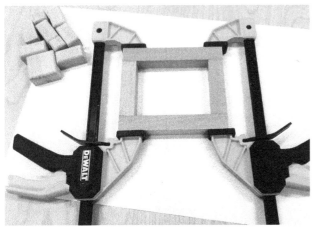

Figure 1-34. Clamping the frame for Dad's Puzzler.

Ideally, you would make 45-degree cuts at the corners of the frame, like a picture frame. But gluing and clamping a 45-degree joint is difficult, so I'm going to postpone it till Chapter 5. Figure 1-34 shows how simpler joints can easily be clamped. This kind of joint, where the end of one piece of wood is glued at 90 degrees, is called a **butt joint**. It's the simplest kind there is.

You still need a base for your tray. Quarter-inch plywood could be used for that purpose, or ABS plastic sheet. I'll be dealing with these materials a bit later. For now, you can glue some cardboard onto the back of your frame.

Because the cardboard contacts the frame over a large area, you can't clamp it easily. Just place a few heavy books on top of the frame to hold it in position while the glue sets. Protect the books from the glue by inserting a sheet of scrap paper. If any of the paper gets stuck in the glue, you can remove it with sandpaper.

My finished version of Dad's Puzzler is shown in Figure 1-35. To get the rounded edges, I had to sand the blocks quite thoroughly.

Figure 1-35. The finished, framed version of Dad's Puzzler, ready to perplex anyone who tries to find the solution.

Solving the Puzzle

Now for the important part. You'll remember that the objective of this game is to shuffle the blocks around so that the large square starts in the top-left corner and finishes in the bottom-left corner. How can this be done?

There's more than one way, but I'll tell you the quickest I have been able to find. In figure 1-36, the top diagram shows the starting position of the blocks, and the bottom diagram shows where they end up. L is the large block, A and B are the two small square blocks, H are horizontal blocks and V are vertical blocks.

In the instructions that follow, the blocks are identified by letters, as shown in Figure 1-36, and an instruction to move up, down, left, or right means that one or more blocks should slide as far as possible in the direction stated. So, "L down" means move the large block as far down as space permits, while "A+B right" means that you move block A and block B together, as far right as space permits.

Each of the horizontal blocks is not identified separately, because there is always only one of them that can be moved in the direction specified. The same applies to the vertical blocks.

Here are the moves:
A+B right, L down, H left, H+B up, A+L right, V up, V+H+H left, A+B down, V+L right, V up, H+H+B left, A up, H right, H down, A+B left, L+H down, H right, V+V+A up, B right, V down, V left, A+B up, V right, V down, A left, B up, H left, H down, A+B right, H up, H left, B down, A+H+H right, V up, V+L left, A+B down, V+H+H right, V up, L+A left, B+H up, H right, L down.

Did you get that? Do you think you can remember how to do it without looking at the instructions?

New kinds of sliding block puzzles are still being invented, and many of them are online. Just Google "sliding block puzzle" and you will find an amazing variety.

But Dad's Puzzler is still my favorite.

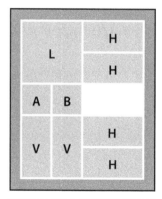

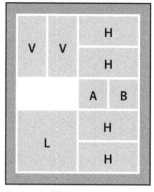

Figure 1-36.
The starting positions (top) and finishing positions (bottom) of the sliding blocks.

Figure 1-37. Even people with a genius IQ may find themselves baffled by Dad's Puzzler.

Handsaw Fact Sheet

Some Saws and Their Teeth

Figures 1-38 through 1-43 show a variety of handsaws, each with its particular advantages.

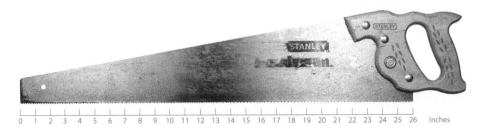

Figure 1-38. A vintage **panel saw** from the 1970s, when handheld circular saws had not entirely taken over the market. The term "panel saw" is derived from its application cutting large panels of wood. This one has nine teeth per inch, although variants were available with eight and 10 teeth per inch.

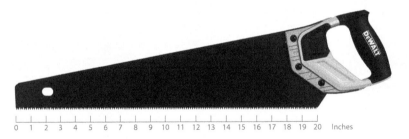

Figure 1-39. A modern panel saw, with nonstick coating and radically improved tooth design. It has eight teeth per inch.

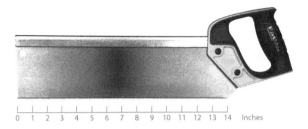

Figure 1-40. A **tenon saw**, also known as a **back** saw or a **miter saw**, with 13 hardened teeth per inch.

Figure 1-41. A general-purpose **utility saw** with 11 teeth per inch.

Figure 1-42. A full-size Japanese-style **pull-saw** with nine teeth per inch.

Figure 1-43. A small Japanese-style **pull-saw** with extremely fine teeth—17 per inch. This is ideal for making clean cuts in soft plastic, as you'll see in Chapter 15.

Figure 1-44 shows the teeth of four of the saws in closeup. A saw with wider teeth generally cuts more quickly and aggressively, but is more likely to leave rough edges. However, the shape of the teeth is even more important.

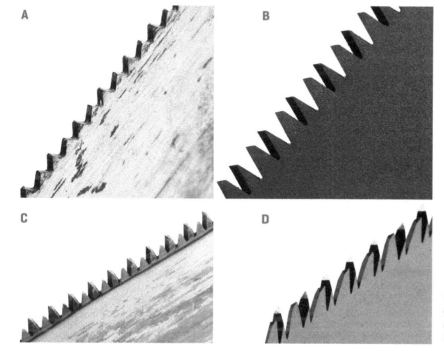

Figure 1-44. Saw teeth look surprisingly different in closeup, and the differences are important. See the explanation on the next page.

The geometry of saw teeth has changed radically over the past couple of decades. Closeup A shows the (rather blunt) teeth of the old 26" panel saw. Closeup B is of the modern panel saw. Closeup C is of the utility saw, in which the darkened teeth indicate that they have been hardened. Closeup D is of the small Japanese-style pull-saw.

Pull or Push?

When using a Western-style saw, you push it to cut the wood. This requires the blade of the saw to be relatively thick, so that it doesn't bend under pressure. Or, in the case of a tenon saw, a thin blade has a thicker bar along the top edge, to stiffen it. Unfortunately, this prevents the full length of the saw from sliding down through a wide piece of wood, and limits the extent of a cut.

A Japanese-style pull-saw cuts when you pull it toward you. Because you are pulling it instead of pushing, the blade can be thin, and its lack of a stiffening strip allows it to penetrate all the way into a cut. Some people feel the Japanese saw is less effective for doing heavy work, and if you pull the saw all the way out of the wood, there may be a slightly greater risk of hurting yourself. Personally I think it's just a matter of what you are used to.

How a Saw Works

Figure 1-45 shows two teeth of a western-style saw being pushed from the right toward the left. Each tooth of the saw is like a tiny knife blade, digging into the surface of the wood and kicking some of it out as sawdust. (In the case of a Japanese-style pull-saw, the blade would be pulled from the left instead of being pushed from the right.)

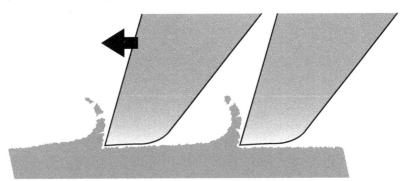

Figure 1-45. Two teeth of a saw, creating sawdust.

If you inspect a saw from the end, you'll see that alternating teeth are angled out to either side. This **set** (the distance off-center) creates the **kerf** (overall cut width) and is shown in Figure 1-46. Because the teeth stick out, they make a cut that is slightly wider than the body of the blade, so that the saw can slide freely.

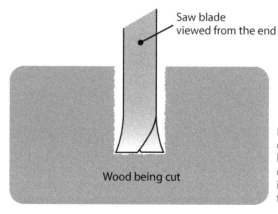

Figure 1-46. End view of a saw blade, showing how the teeth are angled outward to make a cut that is wider than the body of the blade.

Hacksaw

A hacksaw, shown in Figure 1-47, is designed to cut metal. The frame can be adjusted to accept blades 10" or 12" in length, and blades are available with a variety of teeth per inch, 18, 24, and 32 being common. Because the blade is thin and not rigidly clamped, it can flex while the saw is being used. Consequently, precise cuts can be difficult.

Figure 1-47. A typical hacksaw.

Saws That Cut Curves

Various saws that cut holes and curves are discussed in the Holes and Curves Fact Sheet on page 164.

Protecting a Saw

Even a saw that has hardened teeth is easily damaged by banging it carelessly against steel, stone, or concrete. A saw should be placed where it won't be knocked onto the floor accidentally. If you live in a moist environment, protect saws from rust by rubbing oil over them with a rag, or (better) use a wax such as Glidecoat that is formulated for the purpose.

Handsaws vs. Power Saws

For details of power saws, see Chapter 20.

Chapter 2
The Curious Cube

NEW TOPICS IN THIS CHAPTER

- Gluing in 3D

YOU WILL NEED

- Square dowel, ¾" x ¾", length 24"

Also, as listed previously: Tenon saw, miter box, trigger clamps, ruler, speed square, rubber sanding block, work gloves, dust mask (optional), safety glasses (optional), plywood work surface, carpenter's glue, sandpaper, polyurethane, disposable gloves, and paint brush.

Check the Buying Guide on page 248 for information about buying these items.

A Little More Dowel Practice

Let's venture further, now, into three dimensions. This chapter requires no new skills, but it will help you to practice your sawing, clamping, and gluing techniques. And the end product is a toy that has always been one of my favorites.

Known as the Soma cube, it was invented by a Danish scientist named Piet Hein. He created other games and puzzles such as Hex, TacTix, and Nimbi, but the cube was the one that captured people's imagination.

Figure 2-1 shows that you need four pieces of square dowel that are ¾" long (assuming you are using dowel that is ¾" x ¾" square), 10 pieces that are each 1½" long, and one piece measuring 2¼". If you assemble and glue them as shown, you end up with the blocks labeled A through G in Figure 2-2.

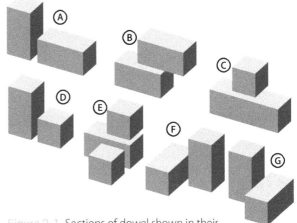

Figure 2-1. Sections of dowel shown in their approximate positions to create seven puzzle pieces.

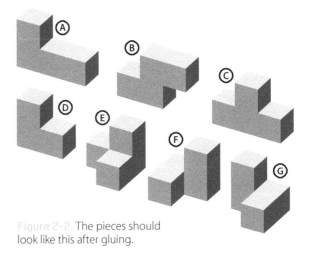

Figure 2-2. The pieces should look like this after gluing.

My version of the puzzle is shown in pieces in Figure 2-3, and assembled to form a cube in Figure 2-4. The assembly process is not too challenging, as there are 240 different ways to do it.

If you search for Soma Cube online, you'll find literally hundreds of other 3D shapes that can be formed with the same pieces.

Figure 2-3.
The seven puzzle pieces.

Figure 2-4.
Can you put together
the pieces to create
this cube?

NEW TOPICS IN THIS CHAPTER

- Hard and soft woods
- Grain and warping
- Sawing without a miter box
- Using a template
- Finishing with paint

YOU WILL NEED

- Awl or pick (may be described as a scratch awl)
- Permanent marker with fine point
- Utility saw (optional)
- Template for drawing holes (see text for explanation)
- Two-by-four pine, no knots, not warped, 9"
- Two-by-four pine, any condition, 12"
- Kilz or similar wood primer, 1 pint
- Latex white paint, 1 pint

Also, as listed previously: Tenon saw, miter box, trigger clamps, ruler, speed square, rubber sanding block, work gloves, dust mask (optional), safety glasses (optional), plywood work surface, sandpaper, disposable gloves, and paint brush.

Check the Buying Guide on page 248 for information about buying these items.

Now that you've worked with square dowels, the purpose of this chapter is to familiarize you with larger pieces of soft wood, such as pine. You'll use it to make two extra-large dice, which sounds simple but will entail techniques of future value.

Lumber Measurements

When wood is sold in the form of beams or boards, it is known as **lumber**. In Canada, Australia, the UK, and some other countries, they call it **timber**; but in the United States, timber is wood that is still inside a tree.

Pine trees are the cheapest source of lumber used in the United States. If you go into a lumber yard or a big-box hardware store and ask for a **two-by-four**, you will usually be offered a piece of pine (sometimes, a piece of fir) about 96" long and approximately 2" by 4" when viewed from the end.

What do I mean by "approximately"? Well, when lumber comes out of the saw mill, it has rough, splintery surfaces, and a two-by-four really does measure 2" x 4". To make it easier to handle, it passes through a **planing machine**, which scrapes off the rough finish and makes the wood smooth to the touch. Because the planer removes about 1/4" from each side, the two-by-four ends up measuring about 1½" x 3½", even though it is still called a two-by-four.

Likewise, a two-by-six measures about 1½" x 5½", a two-by-eight measures about 1½" x 7½", and so on.

What if you buy some one-by-six because you want to build some bookshelves? If it has lost ¼" on each side, like a two-by-four, it should measure ½" x 4½", shouldn't it? Maybe so, but in fact it measures ¾" x 4½". All the "one-by" boards are ¾" thick. See the Wood Fact Sheet on page 33 for more information.

If a two-by-four isn't really 2" x 4", you may be wondering why the dowel that was used in chapters 1 and 2, which was sold as being ¾" square, really was ¾" square. Likewise, if you buy 1" dowel, it really will measure 1" x 1" (allowing for very small quality variations). All I can tell you is, dowels are different. The size on the label of a dowel really is the size of the product.

Because pine is cheap, two-by-fours are usually sold in lengths of 96" (or slightly shorter). If you have difficulty fitting that into a car or carrying it on public transit, the store should be willing to cut it in half for you.

Defects

When using two-by-fours, you have to remember that in the United States, they are mostly intended for framing houses, where the look of the wood is not important. Consequently in a lumber yard you are likely to find some fairly ugly specimens, like the one shown in Figure 3-1.

The oval-shaped dark areas are **knots**, where branches grew out of the tree and died or broke off, leaving the root of the branch to become embedded in the tree. You can try to find better looking wood by asking for "select" grade (which is 80% free from knots), or #1 grade (75% clear), but not all stores carry these grades of wood.

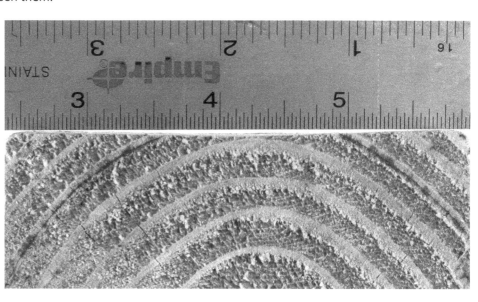

Figure 3-1 Some serious blemishes on a two-by-four.

Knots don't contribute any strength to the wood. For our purposes, you have to cut usable pieces from between them.

To make things more difficult, the wood may have **warped**, meaning that it has curved as a result of the drying process after it was cut from the tree. This can happen across the width of the wood, as in Figure 3-2, which depicts the end view of a two-by-four. The white area under the ruler shows where the wood has warped.

Warping can also happen along the length of the wood, as you will see if you hold it up to your eye and look along it. The take-home message is simple: check two-by-fours carefully before you take them home. For more details, see the Wood Fact Sheet on page 33.

Figure 3-2 The warping of this piece of wood is visible as the thin bright line under the ruler.

Planning the Dice Project

To make two big dice in this project, obviously you will need two wooden cubes. I want to make them out of a pine two-by-four, because pine is one of the softest types of wood and is easy to cut and shape.

Figure 3-3 shows my plan. You'll start with a section of two-by-four which I call the **work piece**. You'll cut a section from the end of it which I call the **dice piece**. Its width will be exactly the same as its height.

The remainder of the work piece is retained as backup, in case you make errors in the next stage.

You will turn the dice piece around and cut two cubes out of it, each cube being the same width as the height of the wood. And there are your dice. Simple!

Actually I can see a couple of snags—but let's get started, and deal with the problems as they occur.

Selecting Your Work Piece

Your two-by-four will have some knots or other defects in it. Select a section of it about 9" long that is clean, smooth, and reasonably flat, and this will be your work piece. In Figure 3-4, you can see the section that I chose. Make a similar selection from your lumber.

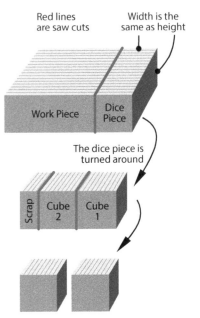

Figure 3-3. The basic plan for cutting two cubes out of a piece of two-by-four.

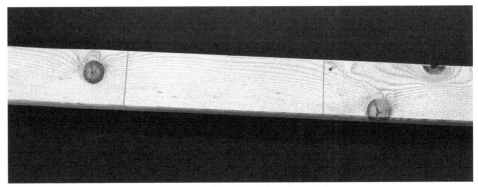

Figure 3-4. Choosing a section of a two-by-four that is free from defects.

When you draw each line across the two-by-four, it's a good idea to extend the line around vertically onto the edges of the wood, as shown in Figure 3-5. This will help you to verify that your saw cuts are square.

Cut the work piece using your miter box, in the same way as you cut the dowel in the previous chapters. If you have a utility saw, it may be quicker than a tenon saw.

If you economized by buying a cheap saw, you may regret this decision while you are cutting the pine. Although pine is much softer than hardwood, the area of a cut in a two-by-four contains more than five square inches, while the area of a ¾" x ¾" dowel is only about ½ a square inch. So, you will be doing 10 times more sawing. In my experience, if your saw has hardened teeth (which I believe are more efficiently contoured), they will reduce the effort greatly. If your saw does not have hardened teeth, you'll just require a bit more patience.

You may find that the extra sawing effort makes the saw more difficult to control. After you have your work piece, check the ends of it carefully. If one of the cuts isn't precisely square, take a look at the other end. You only need one good end. If necessary, you may need to trim another piece from it, to make the cut vertical.

Figure 3-5 Extend each cut line around to the sides of the wood.

You will find some splinters where the saw emerged from the wood, as shown in Figure 3-6. This is a problem with soft wood: it splinters easily. The splinters won't matter in this project, because you will be rounding the edges of the dice. Still, you should bear in mind that splintering occurs when cutting soft wood in a miter box. I'll show you how to avoid it, a little later on.

Cutting the Dice Piece

Smooth all the surfaces of the work piece, using 80-grit sandpaper on your rubber sanding block. To avoid making visible scratches in the soft pine, always rub the sandpaper parallel with the grain. The wood has to feel smooth and flat when you run your fingers over it.

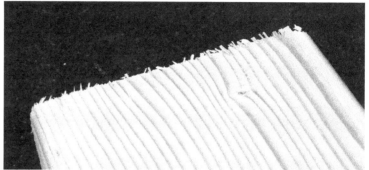

Figure 3-6 A small amount of splintering is typical when sawing pine in a miter box.

Now measure the exact thickness of your work piece. As you see in Figure 3-7, my two-by-four was slightly less than 1½" thick after I finished sanding it. In fact, it was almost exactly 1⁷⁄₁₆". You can count the divisions to the right of numeral 1 on the ruler in the photograph, and you'll see there are seven of them. Bear in mind that ⁸⁄₁₆" is the same as ½", so ⁷⁄₁₆" is one sixteenth less than ½".

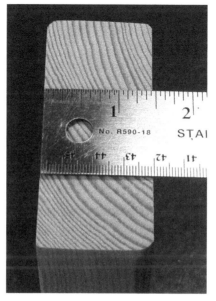

Figure 3-7 Measuring the exact thickness of the wood.

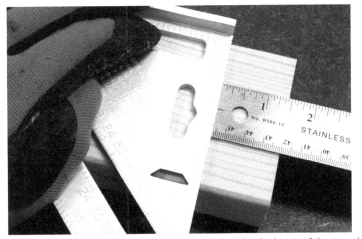

Figure 3-8. Applying the measurement, equal to the thickness of the wood.

Figure 3-9. Drawing the cut line.

Whatever the thickness is of your two-by-four, this will be the size of your dice. In Figure 3-8, you see me applying this measurement to the top surface of the work piece, and in Figure 3-9 I am drawing the line. I suggest you should extend this line around the sides of the wood, as you did when cutting the work piece, because everything has to be precisely square in this project.

Cut the dice piece in your miter box, and check its size and squareness carefully before proceeding.

Now, how are you going to extract the two cubes from the dice piece? Uh-oh. We have a problem here. As you see in Figure 3-10, the dice piece is relatively small. You're going to have a hard time holding it securely in the miter box. And even if you succeed, you will be left with an even smaller piece, from which you have to obtain the second cube.

The time has come to think outside the box. The miter box, I mean. You will have to clamp the wood directly, and saw it freehand.

Figure 3-10. The dice piece in the miter box will be difficult to hold.

Outside the Miter Box

Here's how it's going to work. In Figure 3-11, you see the dice piece has been inserted between two new pieces of wood, identified as a **guide piece** and a **sacrificial piece**. Everything will be clamped together while you saw along the red line.

The saw will slide against the edge of the guide piece, which will help you to make a good vertical cut. When the saw cuts down through the dice piece, it goes into the sacrificial piece, which prevents splintering. Of course, the saw will make a mark in the sacrificial piece, but that's what it's there for. It sacrifices itself so that the piece above it has a nice clean edge.

Using a sacrificial piece is a common technique to avoid splintering.

The sacrificial piece is just any section of two-by-four. It can have knots or other blemishes. Its condition is not important, so long as the surface is reasonably flat. It should be 8" to 12" long.

You can cut the guide piece from the remainder of your work piece. It should be about 2" long. Its edge should be absolutely, precisely vertical, so use the miter box to cut it, and check it carefully afterward.

Do you think this will work? Well, it worked for me, as you will see in the next photographs.

In Figure 3-12 the dice piece has been marked at a distance from the end equal to the thickness of the wood.

In Figure 3-13 the pieces have been stacked up, ready for cutting. Two clamps are necessary, to prevent the wood from rotating in response to the pushing action of the saw. There's just enough room for the clamps, side by side.

Note the gap between the guide piece and the pencil line, to allow room for the thickness of the saw blade. The saw must not cut into the line. Also, be careful that the saw does not rub against the metal bar of the clamp.

Saw through the dice piece, keeping the saw blade flush against the guide piece. You are cutting with the grain, now, instead of across the grain. The saw feels different; it tends to skate along the wood instead of digging in. You can buy saws that are specifically designed for this. A **rip saw** cuts with the grain, but is too big for our purposes. A regular saw will work.

All the pieces are clamped together while being cut.

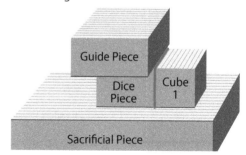

Figure 3-11. A strategy for cutting cubes out of the dice piece.

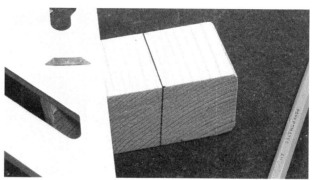

Figure 3-12. The distance between the pencil line and the end of the wood is equal to the thickness of the wood.

Figure 3-13. Stacked and ready!

Make your cut, and now you have a cube, as in Figure 3-14. It has two rounded edges, but that's okay, because you are going to be rounding all the edges.

There's just enough of the dice piece left to allow you to remove a second cube. But you'll have to turn the piece around. The section that you clamp will become your second cube, while the right-hand end is going to be discarded. This means you must now cut to the right of your pencil line, not the left. The pencil line must disappear under the guide piece, as shown in Figure 3-15.

Figure 3-16 shows the cut completed, and 3-17 shows the two cubes.

Figure 3-14. First cube extracted from the work piece.

Figure 3-15. Ready to cut the second cube.

Figure 3-16. The last cut, completed.

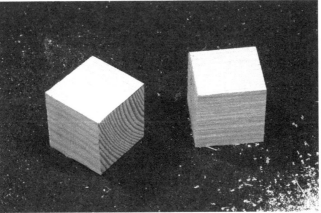

Figure 3-17. Two cubes, ready for smoothing and shaping.

Flat Sanding

Run your fingers lightly over the cubes that you have cut, and you'll find some imperfections. Probably the saw left some marks, and the grain of the wood creates small ripples in the surface. It's time to make everything look good.

Because you want the surfaces of your cubes to be absolutely flat, I suggest you clamp a piece of sandpaper to your flat work surface, as shown in Figure 3-18.

Rub the faces of the cubes over the sandpaper until they feel really smooth. Now check that all the sides are the same dimensions. You may have to do more sanding to make everything the same.

I used 80-grit sandpaper to get a quick result. Always remember to sand parallel with the grain of the wood, to avoid creating visible scratches. Also sand away from the clamps, as the sandpaper will tend to buckle if you try to sand toward them.

Because pine is one of the softest woods, finishing your dice by hand shouldn't take too long.

Rounding the Edges

When you have two acceptable cubes, you're ready to round their edges. This is necessary for the dice to roll easily. Just drag each cube across the sandpaper on the workbench, while turning the wood while you drag it, to make a curve along each edge.

The result should look like Figure 3-19.

Lastly, round the corners. I think this is easiest while holding a sanding block in one hand and the cube in your other hand. Use a rotating motion of your wrist, and don't press too hard. The fully shaped dice are shown in Figure 3-20.

Painting

Your last step is to mark the dice with spots. This creates a problem, which is that two faces of each cube show the ends of the wood grain, while the remaining four faces show the sides of the grain. The ends and the sides look significantly different from each other, and you may want your dice to look the same from any angle.

The answer to this problem is . . . paint!

Figure 3-18. To sand flat surfaces, the sandpaper must be flat. Here's one way to achieve this.

Figure 3-19. The cubes with rounded edges.

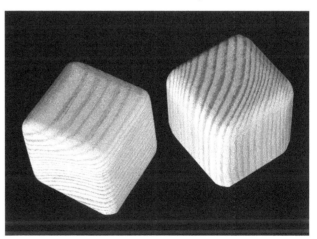

Figure 3-20. Fully shaped and ready for finishing.

Figure 3-21. One way to support the dice while you put primer on the wood.

Figure 3-22. Applying primer with a paper towel.

I don't enjoy painting, because it takes too long and never looks quite as perfect as I would prefer. Raw pine is especially annoying, because the wood is so absorbent, you have to begin with a primer that seals the wood.

My favorite primer is called Kilz. Probably the name comes from the fact that it kills the tendency of stains on wood to bleed up through paint and become visible. I like the oil-based version of Kilz, but the Premium Latex version also has good reviews, and allows easier cleanup.

How should you hold the dice while you paint them? I decided to make a support for them by cutting two triangular holes in a piece of corrugated cardboard, as shown in Figure 3-21. Each triangle holds three faces of a cube while allowing access to the other three, after which you can turn the cubes over and paint the other faces. Each coat of Kilz takes about one hour to dry.

You can use a brush to apply primer, but for a small job like this I prefer paper towels or a rag, because you won't get any brush strokes. Naturally you need to wear gloves. I suggest the same disposable gloves that I mentioned when you were applying polyurethane in Chapter 1. Painting in progress is shown in Figure 3-22.

Your top coat can be of whatever paint you choose. I used a semigloss latex white. You may have to apply two coats to get a uniform result, and once again I suggest applying it by dabbing a paper towel.

Lastly, the Spots

It's easier to draw with precision on a piece of paper than on a wooden cube. Therefore I suggest drawing a layout for the centers of the dice spots on paper before transferring them to each face of the cube.

The left half of Figure 3-23 shows how I think the spots should look on the dice when they display a number five. The right half of Figure 3-23 shows a grid that you can draw to guide you in creating the spot layout.

Each spot will be ¼" diameter.

Figure 3-23. The spot layout for the dice, and the grid that will help you to create it. Draw the grid on a piece of paper.

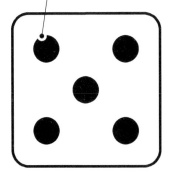

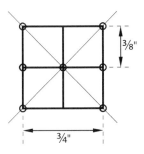

After you draw the grid on paper, poke a hole in each of the seven locations that are circled. To do this, you can use a handy tool known as an **awl** (also known as a **scratch awl**). It looks like a small screwdriver, but terminates in a point, as shown in Figure 3-24.

You'll be using the awl in many future projects. It's more versatile than it looks.

Alternatively you can use a **pick**, which is a skinny version of an awl, and may be a little easier to control with precision. Pick-and-hook sets such as the one in Figure 3-25 are sold very cheaply, and the hook-shaped tools may be useful for retrieving parts that fall into small places. Personally I prefer an awl, because it stands up to heavy treatment—for instance, if you want to punch a hole in a piece of drywall, which will be necessary in Chapter 12.

After pricking holes through each dot, center your drawing very carefully on the face of one of your dice, as shown in Figure 3-26. Use a fine-point Sharpie marker to poke through some of the holes, corresponding with the spot pattern that you want to create. Remove the template, and now you need to enlarge each dot to ¼" diameter, or whatever you think looks appropriate.

There are two ways to draw circles for the spots. The precise way is to use a plastic guide of the kind shown in Figure 3-27, which is sold as a **template** in stationery stores. If you don't want to spend money on that, use any object with an appropriate circular hole in it, such as a door key.

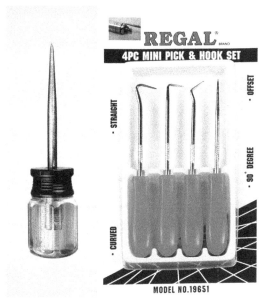

Figure 3-24. An awl, also known as a scratch awl.

Figure 3-25. A bargain-basement pick-and-hook set from a rural convenience store. The pick is the one that doesn't bend at the end.

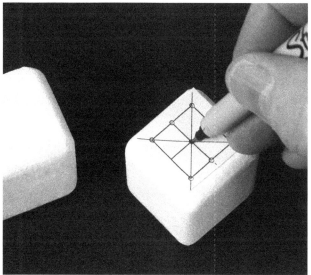

Figure 3-26. Center your drawing on the face of one of your dice, and use a fine-point marker to poke through some of the holes that you made.

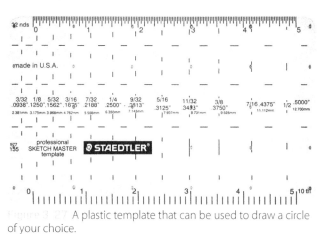

Figure 3-27. A plastic template that can be used to draw a circle of your choice.

Is there a way to make spots on the dice using black paint? Yes, if you are highly skilled with a tiny paintbrush. Personally, I think a permanent marker is good enough. Maybe there is some other way that I haven't thought of.

My finished dice are shown in Figure 3-28. Incidentally, when you are drawing the spot patterns on them, conventional dice always put patterns that add up to 7 on opposite sides. If you are not a conventional person, maybe you don't care about this. I just thought I'd mention it.

Figure 3-28. The finished dice.

One More Project

The dice in this project were big, but how would you like to make dice that are absolutely huge? Perhaps twice as big in each dimension?

This wouldn't be so hard. Begin with two pieces of two-by-four, and glue them together, face-to-face.

The gluing won't work properly unless the surfaces are exactly flat, which will entail some fairly relentless sanding. Then use carpenter's glue, clamp them with as much force as you can, and leave them overnight. You'll now have a piece of wood that is 3" x 3½" (because each of your two-by-fours was 1½" x 3½"). You'll have to slice the side off it before you cut two cubes out of it.

Not trivial—but doable!

Wood Fact Sheet

Beams and Boards

Lumber is wood that is extracted from trees by a saw mill, to create **beams** and **boards**. In the United States, a beam is typically a "two-by," such as a two-by-four or two-by-eight. A board is typically a "one-by," such as a one-by-six or a one-by-ten.

Two-by beams measure approximately 1½" thick, because the rough-sawn wood has been passed through a planing machine. One-by boards measure approximately ¾" thick. Beams and boards all measure ½" less in width than their name suggests; thus, a two-by-four is 1½" x 3½".

Most houses in North America are framed in wood. The walls are built around two-by-fours or two-by-sixes. Traditionally, two-by-eights or taller beams were used underneath the floors, to support them, while the floors themselves were made out of boards. These beams and boards are now being replaced in various ways by composites such as plywood or OSB. See the Composites Fact Sheet on page 78 for more information.

When a tree is cut down, the exposed end of it shows rings that formed as a result of the tree growing faster in the summer and slower in the winter. When the tree is sawn into planks, the rings are elongated and appear as grain. See Figure 3-29.

Figure 3-29. Growth rings and grain are two views of the same thing.

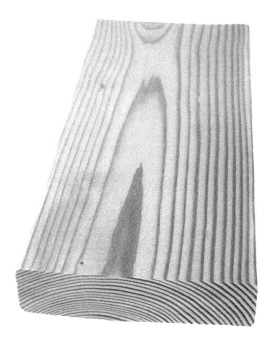

Figure 3-30. How a tree trunk is cut into boards and beams when it is cut "through and through."

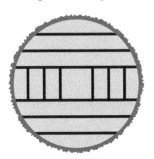

Figure 3-31. Another common method for slicing up a tree trunk. This is known as "plainsawn."

Figure 3-32. Another common method for slicing up a tree trunk. This is known as "quartersawn."

How Wood is Removed From Logs

The simplest, quickest, and most economical way to saw lumber out of a tree trunk is to make parallel cuts, as in Figure 3-30. This pattern is known as **through and through**. In another common approach, the wood is **plainsawn**, as in Figure 3-31, allowing a square piece to be removed from the middle. (Posts that measure 4" x 4" or larger are often used in construction work.)

Quartersawn wood can be cut as shown in Figure 3-32, the advantage being that the grain is now more symmetrical across each board. The discourages warping, as explained below. However, this is the most wasteful and expensive way to remove lumber from a tree.

Warping, Cracking, and Other Problems

A tree in a forest contains a lot of water. After the tree has been felled and cut into sections, water evaporates from the surface of the wood until the concentration inside the wood is in balance with moisture in the air.

Wood does not lose water evenly, and it shrinks as it dries. As some areas shrink more than others, this creates stresses which cause the wood to warp, meaning that it bends.

If you look at the growth rings at the end of a piece of lumber that has been plainsawn or cut through-and-through, the pale rings may have shrunk more than the dark rings, causing the wood to warp across its width. This is known as radial shrinking. Quartersawn lumber is much less likely to suffer from this problem.

Wood can also crack or split if the outer layers dry too rapidly and shrink relative to the inner layers. Cracking is most likely at the ends of a beam, but you should look for splits along the whole length of it, too. Figure 3-33 shows a two-by-eight that split, probably because of rapid drying.

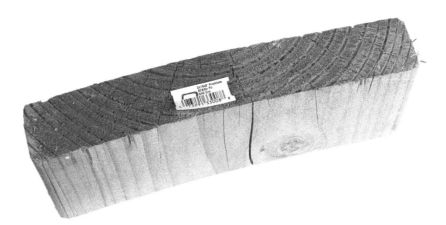

Figure 3-33. Splits tend to develop at the ends of lumber when it is dried too quickly, regardless of the "premium" classification on the label.

Figure 3-34 shows how wood can lose some of its moisture as resin. Eventually the resin dries and can be scraped or sanded off, allowing the wood to be used.

Choosing Wood

Some lumber is **air-dried** in stacks outside the saw mill, but most commercial products are **kiln dried** in buildings like giant ovens. Kiln drying can be more accurately controlled, and allows more moisture to be removed. It is also faster than air drying.

Figure 3-34. Beads of resin oozed out of this piece of word dried during the six months after it was purchased.

Either way, the drying process may be incomplete when wood shows up for sale. If you see a fresh stack of two-by-fours in a lumber yard or big-box hardware store, remove a few pieces from the top of the stack and run your fingers over a piece that has been underneath. Wood that has not dried sufficiently will actually feel moist to the touch, and you can almost guarantee that it will warp after you buy it.

Even when you find wood that seems to be well dried, it's still a good idea to store it yourself for a few months, if possible, to give it time to equalize with the surrounding air.

Hard and Soft Woods

White pine, **poplar**, **douglas fir**, **cedar**, **maple**, **birch**, and **oak** are some commonly available woods, listed here in sequence from soft to hard (although variants may differ). A sample piece of red oak board is shown in Figure 3-35. Note how close-spaced and uniform the grain is—completely different from pine.

Figure 3-35. A nice piece of red oak board.

Generally speaking, those described as soft woods really are softer and easier to cut and shape, but they are properly defined as coming from evergreen trees. Soft woods tend to be paler in color than hardwoods, and because they grow faster, they are usually cheaper. The faster growth creates rings and lines of grain that are more widely spaced and more easily visible than the grain in hardwoods.

Hardwoods come from deciduous trees—that is, trees that have leaves as opposed to spines, and mostly shed their leaves in the winter. Oak is probably the most widely known hardwood, is extremely heavy, and is difficult to work with. If you are trying to round the edges of a piece of oak using handheld sandpaper, it's going to take a while.

The greater density of hardwood, and the close spacing of the growth rings, help to reduce the problem of warping. Even so, you should check each length before buying it. Don't fall into the trap of thinking that because hardwood is much more expensive than soft wood, it will never create problems for you.

A general lumber yard does not usually store a wide range of woods, but in many towns, specialty stores offer a greater variety. Their wood may cost more, but is likely to be of higher quality.

Clear and Common

Wood described as "common" has knots in it. A knot is a blemish where a branch grew out of the trunk of a tree, then died and became embedded in new growth.

Knots cause several problems:
- They have no structural strength. They are almost equivalent to holes in the wood. In fact, sometimes a knot shrinks and falls out, leaving a hole.
- If a knot pops out of a piece of wood while you are using a power saw, this can disrupt the cut or create a hazard.
- Knots are denser than the rest of the wood. They are difficult to saw, and you may not be able to hammer nails into them. They will not grip the thread on a screw.
- Many people feel that knots are ugly, especially in objects such as furniture.
- If you are painting some wood, knots will absorb much more paint than the surface of the wood. Really you should sand and seal them first, using a compound for that purpose. Even then, they will still tend to be visible under the paint.

If a length of wood is taken from deeper inside the tree, it can be free from knots. This is known as "clear" lumber. Unfortunatelty, it costs two or three times as much as common lumber. Figure 3-36 shows two pine boards, the one at the top being common, while the one at the bottom is clear.

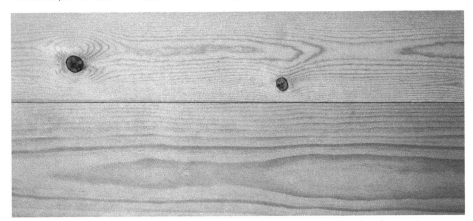

Figure 3-36. The difference between common lumber (top) and clear lumber (bottom). Unfortunately, there is also a substantial price difference.

Chapter 4
Nailing It

In this project, you'll create a randomizer by hammering nails into a board. What do I mean by a randomizer? Well, imagine a lot of nails standing like trees in a forest on a steep slope. If you roll a marble among them, it should bounce around and come out at the end with an equal, random chance of being on the left side or the right side. If you label the left side "yes" and the right side "no," you can use the randomizer in games, or to make difficult decisions, or just for fun.

You can also use it to demonstrate some trickier aspects of probability, which I'll describe at the end.

Previously I mentioned that one-by-six pine is actually ¾" x 5½". When you go shopping for it, probably you will be expected to buy a piece 96" long, like the two-by-four that you were using previously. You only need 12" of one-by-six for this project, but you'll be using more of it later in the book, and it won't cost very much. I think 96" will be worthwhile.

The **hammer** can be the cheapest you can find. If you are physically small or very young, you can consider choosing an 8-ounce hammer, bearing in mind that the nails in this project are small and don't require a lot of force. 10-ounce and 12-ounce hammers may also be found. Figure 4-1 compares an eight-ounce and 16-ounce hammer.

NEW TOPICS IN THIS CHAPTER

- How to drive nails
- Inserting nails to a uniform depth
- Drawing a pattern
- Types of nails
- Types of hammers

YOU WILL NEED

- Claw hammer, between 8 ounces and 16 ounces (see text for discussion)
- Pliers, long-nosed preferred, generic slip-joint acceptable
- Two-by-four pine, any condition, 12"
- One-by-six pine, no knots, not warped, 12"
- Square dowel, ¾" x ¾", length 3"
- Finishing nails, 1¼", 1 lb box
- Marbles, ⁹⁄₁₆" diameter (14mm), quantity 20
- Half-inch masking tape

Also, as listed previously: Tenon saw, utility saw (optional), awl, trigger clamps, ruler, speed square, rubber sanding block, work gloves, dust mask (optional), safety glasses (optional), plywood work surface, carpenter's glue, and sandpaper.

Check the Buying Guide on page 248 for information about buying these items.

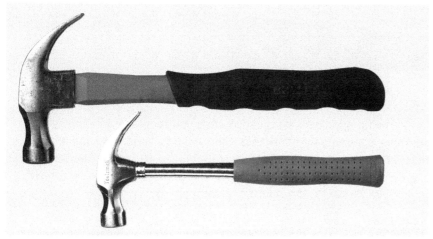

Figure 4-1. Two weights of hammer.

Long-nosed pliers are the easiest type to use in this project, as shown in Figure 4-2. If you only have **generic slip-joint pliers** like those in Figure 4-3, you can use them instead.

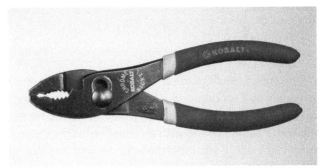

Figure 4-2. Long-nosed pliers, about 5" in length, are useful for holding nails while you are hammering them.

Figure 4-3. Generic slip-joint pliers.

To learn what a **finishing nail** is, and a lot more facts about nails, see the Hammer and Nail Fact Sheet on page 48. A box of 1¼" finishing nails will probably look like the one in Figure 4-4.

For the marbles, if you don't have any, you can buy them super-cheaply from Walmart or eBay. The size must be ⁹⁄₁₆" diameter (or 14mm), to match the spacing between the nails.

Figure 4-4. A typical 1lb box of finishing nails.

Safety While Hammering

Before you start hammering nails, consider whether you should be wearing **eye protection**. There is the very slight risk that when you hit a large nail, a metal fragment could break loose and fly off at dangerous speed. Personally I have never seen this happen, and in any case, the projects in this book only require small nails. I'm just mentioning the safety risk for future reference.

Hitting your thumb is a more realistic risk, because everyone does that sooner or later. You can eliminate the risk by using pliers to hold each nail, but after a while you may get tired of doing this.

Hammering Practice

This project requires accurate nail spacing, so it's a good idea to get in a little practice. You can use a scrap piece of two-by-four to test your hammering skills. First check Figure 4-5 to make sure you have the basic concept figured out.

Starting the nail is easy. With your left hand (if you are right-handed), use the pliers to hold the point of the nail against the location where it's going to go. Hold the nail vertically, hold the hammer near its head (although not so close that you can bang your fingers against the nail), and give the nail two or three small, gentle taps.

When the nail is able to stand up straight on its own, take the pliers away. Shift your grip toward the end of the handle of the hammer, and whack the nail a few times, being careful to bring the head of the hammer straight down. Holding the end of the handle allows you to apply greater force, because the hammer is like an extension of your arm, multiplying its leverage.

However, when you are using it this way, the hammer becomes more difficult to control. If you find that you aren't hitting the nails on-target, shift your grip closer to the head of the hammer. Your strokes will be less powerful, but you can make them more accurate.

What if a nail starts to go in at an angle? Straighten it immediately with your pliers. And if a nail gets bent? Pull it out with the pliers, and throw it away.

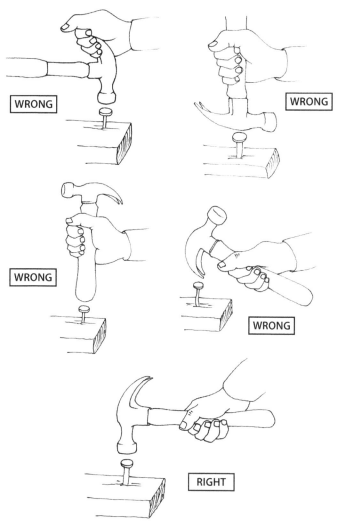

Figure 4-5. Five ways to hammer in a nail.

Nail History

Long ago, in colonial America, nails used to be made by hand. They were so precious that if a house was abandoned, people would burn it down just to retrieve the nails. Thomas Jefferson derived some income by owning a blacksmith business that made and sold nails.

These days, nails are so cheap, we don't need to waste time trying to straighten them.

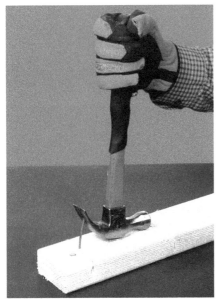

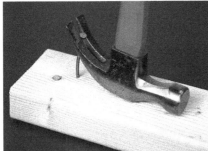

Figure 4-6. Using a claw hammer to extract a nail.

A **claw hammer** is designed so that you can use it to pull out a nail, as shown in Figure 4-6. In this project, the nails will be spaced too closely to allow this, and pliers are the only option. The nails are so small, you should be able to pull them out without much trouble.

Splitting Wood

If you hammer a nail too close to the end of a piece of wood, the force of the nail can break the wood apart. The risk increases if the nail is too thick, or too close to the end of a piece of wood. Hardwood may split more easily than softwood. Thin wood splits more easily than thick wood. And if you insert two or more nails in line with one line of the grain, the wood is more likely to split.

My successful attempt to split a piece of pine is shown in Figure 4-7. I had to use a relatively thick nail to do it. I had a much easier time splitting particle board, as shown in Figure 4-8.

It's a good idea to see if you can split a piece of scrap wood, just so that you learn the limits. Using the little 1¼" nails that I specified for this project, I doubt you will be able to split a piece of pine—but you can probably split some ¾" square dowel, especially if it is hardwood, which is less willing to make room for the nail.

Figure 4-7. How to split a piece of pine.

Figure 4-8. Particle board splits very easily.

Cutting the One-by-Six

You will need a 12" section of one-by-six pine that has no knots in it. Knots are unacceptable for this project, because they are hard and brittle, so you won't be able to hammer nails into them.

How should you go about cutting your section of one-by-six? Bear in mind a one-by-six is too wide to fit in your miter box. If you had a panel saw and a sawhorse, that would probably be the most obvious way, but I am assuming that you are working on a table top with a tenon saw.

With this in mind, Figure 4-9 shows my suggestion. You will draw a guide line across the board with your speed square. Underneath the board you will put a sacrificial two-by-four which is 5½" long— the same length as the width of the board. (You can cut the sacrificial piece in your miter box.)

Figure 4-9. How to cut a piece of one-by-six board, which won't fit in a miter box.

Place the sacrificial piece so that it extends out over the edge of your work area, as shown in the figure. This will allow room for two clamps, and for you to make a saw cut. The figure shows a cut in progress, although I took the saw out of the photograph so that you can see the setup more easily.

Notice another spare piece of two-by-four that I tucked under rear end of the board, just to keep it level.

If you feel you need a guide piece to keep your cut straight and vertical, as in the previous project, you could add another 5½" section of two-by-four for this purpose on top of the board, and extend the clamps upward to hold it. But the precision of this cut is not crucial, and I think you should try making it without a guide piece, just to test your sawing skill.

You don't even need the sacrificial piece, if you don't care about splinters. But that's up to you.

When you have your 12" of knot-free one-by-six board, you need to make sure it is smooth, so that marbles will roll around on it freely. Sand the board until you cannot feel any ridges of the grain when you run your fingers over the surface.

Making a Plan

For this project, the array of nails has to be in a triangular pattern, but I figured out a simple way to draw it from a grid of rectangles. All you need is a letter-sized sheet of paper, a ruler, and preferably some pens in three colors.

Begin with the grid shown in Figure 4-10. I've shown the dimensions in millimeters, because they are a lot easier to deal with than sixteenths of an inch. Inches are mandatory in the workshop environment in the United States, but there's no need to make things more difficult than necessary when you're just working with a pen and paper.

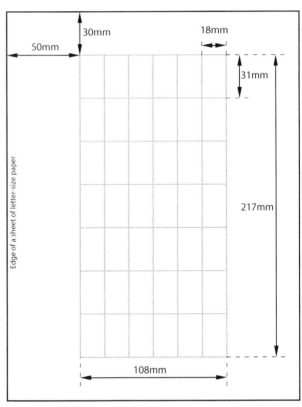

Figure 4-10. The first step in drawing a grid to position nails in this project.

I've drawn the rectangles in green, because I'm going to add other colors.

Now using black ink, draw the diagonal lines shown in Figure 4-11, connecting the corners of the rectangles. Try to make the lines accurate.

Make red dots where the black lines cross, as shown in Figure 4-12. These will be the positions where you hammer in the nails. Their positions are important, because the marbles will roll among them. Extra nails will be inserted on the black dots, and their positions are not so important, because their purpose is just to stop the marbles from rolling out and escaping from the grid.

How will you transfer this pattern onto the wood? I'll show you as soon as I deal with some necessary setup.

Precision Hammering

One of the most important factors, when you are hammering nails, is to have a firm support. If you rest the board in the center of a table, the table will tend to vibrate. This means that a lot of your energy will be wasted, causing the table to jump around.

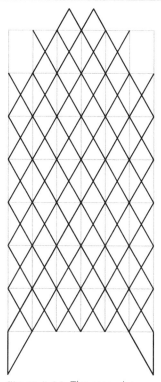

The best support would be a very solid workbench, but if you only have a table, position your work over a leg of the table, to transmit your hammering force directly to the floor. I am assuming that you are using a wooden-topped table, not a glass-topped table!

Sit close to the work. If you were hammering 3" nails, you might want to stand back and apply some body weight to the hammer. For this little job, that isn't necessary. Position yourself close to the nails so that you can see them clearly.

Tape your plan to the piece of two-by-six, and use an awl before you start hammering. This sounds like an unnecessary extra step, but it will save you time. Position the sharp point of the awl exactly where you want each nail to go. Make sure the awl is vertical, and push down through the paper to make a prick mark in the wood. Now when you hammer in the nail, it will position itself automatically on the mark, and your job will be much easier. When I built this project, I began by making prick marks with the awl for all of the nail locations.

Figure 4-11. The second step in creating a grid.

Hammer the nails through the paper into the wood, as shown in Figure 4-13. When you finish hammering, you can rip the paper away. This eliminates the need to draw lines on the wood and erase them later.

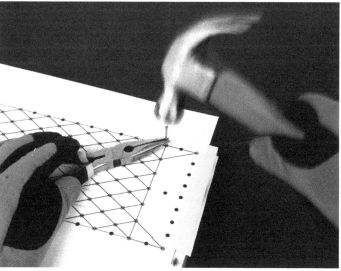

Figure 4-13. Hammering the first nail. The paper is attached to the pine board by some green masking tape.

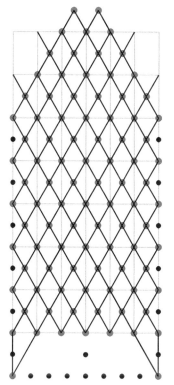

Figure 4-12. Nails will be located where you see red dots and black dots.

Depth Control

If you hammer in a nail too far, it will go all the way through the board and emerge underneath. To avoid this, you need to do some measuring.

Bearing in mind that the board is ¾" thick, and each nail is 1¼" long, how deep should you hammer each nail?

It will be easier to figure this out if you convert everything to eighths of an inch. The ¾" board is ⁶⁄₈" thick. Each nail is ⁵⁄₄" which is ¹⁰⁄₈" long.

To make sure the nails don't go all the way through, you can hammer them ⁵⁄₈" into the wood. That will leave a ⅛" margin of error.

Because each nail is ¹⁰⁄₈", the exposed part will also be ⁵⁄₈". When you start hammering, you can use a ruler to check how much of each nail is sticking out, and Figure 4-14 shows me doing this. In this figure, you can see that ¾" of the nail is exposed. But ¾" is the same as ⁶⁄₈". So I can hammer it in another ⅛".

Figure 4-14. Measuring the height of a nail.

In Figure 4-15 I added a second nail at the other end of the diagonal line where the first nail is located, and stretched a piece of yarn between them. Now the nails along that line can be hammered in so that they are all at the same height as the yarn.

Stretching a string is a method often used on construction sites—for example, to make sure each layer of bricks in a wall is properly aligned. The first row of nails is shown in Figure 4-16.

Some people enjoy being precise. Others don't. If you don't, you can skip the figuring, and complete this project just by guessing the depth of each nail and laying your ruler along them once in a while to verify that they are about right. You will run a greater risk of a nail going all the way through and poking out at the bottom, but maybe that won't be important to you. The main thing is for you to complete the project in a way that satisfies you.

Rolling Randomly

When your nailing work is complete, it should look like Figure 4-17. There are about 120 nails; how long do you think you will take to hammer them all in? Suppose it takes you 15 seconds per nail. That would add up to about 30 minutes. But I think you'll get quicker as you proceed. The project is not as challenging as it looks (so long as you use the awl to locate each nail properly).

Now you can rip the paper away. Tilt the wood upward at a 45-degree angle, and drop a marble in at the top. Can you predict where it's going to end up, as it bounces among the nails? I think not. That's why this little gadget is called a randomizer: it is unpredictable. Try dumping several marbles at once, and see where they go, as in Figure 4-18.

One thing you'll notice is that some marbles get stuck at the edges of the board. In many of the projects in this book, I'm going to leave you room to make improvements. How would you prevent the marbles from lodging like this? Could you add a few extra nails to take care of it?

Here's another question. What if you try to cheat by dropping a marble a little to one side, between the topmost nails? This might affect the outcome, so maybe you should take a precaution against it. You could add a couple more nails above the ones at the top, and space them the precise width of a marble. Now it should fall exactly the same way each time— or will it?

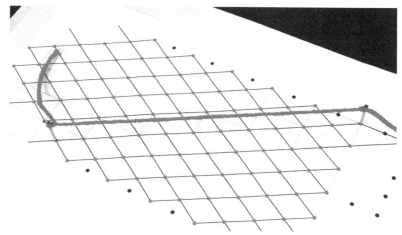
Figure 4-15. Intermediate nails can be hammered in so that they are the same height as the yarn.

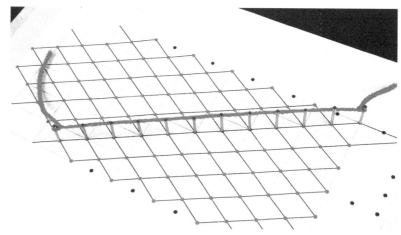
Figure 4-16. The first row of nails.

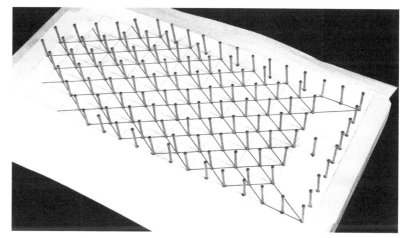
Figure 4-17. The completed field of nails. With more than 100 of them, you should have acquired some excellent hammering skills by the end of this project.

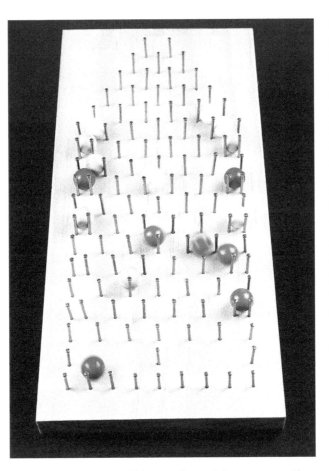

Figure 4-18. A test run shows how widely the marbles can disperse, even though they all enter the field at the same place.

I think you will find that tiny variations are still enough to have an effect. The marble may have a little spin on it, or there may be a current of air, or some other factor. In science, factors of this kind are known as *uncontrolled variables*.

Suppose the left area at the bottom is labeled "yes" and the right area is labeled "no." If this is a really random device, and you test it many times, you should expect a marble to end up an equal number of times in the "yes" and "no" areas. You can try to verify this, perhaps using multiple marbles to speed things up. If the outcome is equally likely to be "yes" or "no," we say that the options are *evenly weighted*.

Do you think you would have affected the outcome if you spaced the nails farther apart? The marbles might bounce around more, making their exit point less predictable. Or would they?

What if you tilt the wood more or less? Will that affect the randomness of the outcome?

The Bean Machine

I designed the randomizer as a smaller version of a device sometimes known as a *bean machine*. An example is shown in Figure 4-19. It contains dozens or hundreds of small, hard objects, such as beans, which can be dumped into a reservoir at the top. Imagine that a pane of glass, in the front, prevents the beans from falling out. When they trickle down among the pins, they stack up in a heap shaped like the one in the figure. More beans accumulate in the middle than at the ends because there are more possible paths that the beans can take to the middle.

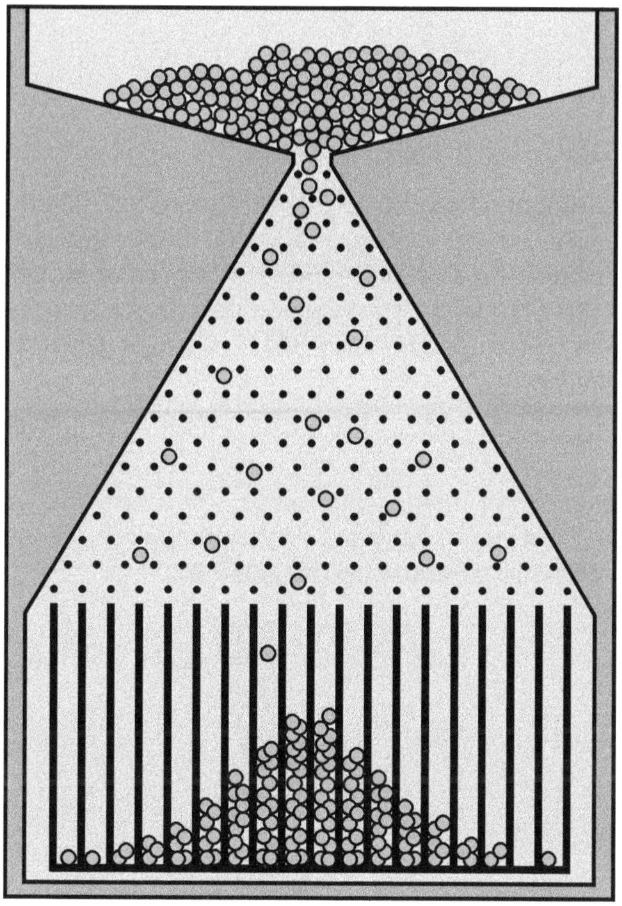

This is an important concept. If you look at the extreme right-hand end of the reservoir at the bottom, there is only one path a bean can take to end up there. But by my count, there are 17 possible paths leading to the next position along, and there are probably hundreds of different paths leading down through the nails to the center positions.

Incidentally, "bean machine" is a widely used name for this kind of gadget. It's not just something that I made up. It even has a Wikipedia entry. You can check it out.

The profile of the stacked beans at the bottom is often known as a *bell curve*, because it is vaguely shaped like a bell. This curve crops up in many areas of statistics. SAT scores, for instance, tend to be distributed in a bell curve, with most people scoring near the middle and only a few at the extremes. Other examples are inaccuracies in production processes, variations in value of a blue-chip stock over a long period, and some astronomical phenomena.

The formal name for a bell curve is *normal distribution*, which is a big concept in the social sciences, and also in mathematics. If you had enough patience, you could make a bean machine to model normal distribution fairly accurately. But why does it turn up in so many different places? That's something you'll have to read about on your own. Meanwhile, have fun testing the randomness of your randomizer.

Figure 4-19. In a bean machine, the beans accumulate in a bell curve.

Hammer and Nail Fact Sheet

Hammer Weights and Types

The **weight rating** of a hammer is the weight of its head only. An average finishing hammer has a 16-ounce head and a 16" handle. Smaller, lighter hammers are appropriate for smaller nails that don't require so much driving force, and for smaller people who may have less physical strength. A heavier hammer requires more strength, but its greater momentum can sink larger nails more quickly—if you can control it.

At least a dozen different styles of hammers exist, the **claw hammer** being the most popular variety. The nail-pulling action of the claw provides a lot of leverage, but this force is transmitted downward by the head of the hammer, and can damage the wood that you're working on. To prevent this damage, insert a small piece of plywood or other thin scrap material. See Figure 4-20.

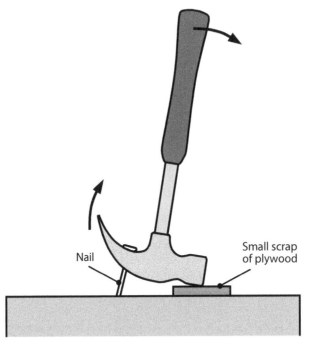

Nail

Small scrap of plywood

Figure 4-20. How to prevent damage to a wood surface by inserting a small scrap of plywood when using a claw hammer to remove a nail.

All hammers used to have **wooden handles**, and some traditionalists still prefer their look and feel. However, the wood tends to loosen in the head of the hammer, and I see no reason to use this type of hammer anymore. A **fiberglass handle** is lighter and stronger. You can also buy a hammer with a **steel handle**, which is the strongest type, but it adds weight where you don't really want it. Weight concentrated in the head of the hammer is more useful, because that's the part that hits the nail.

A hammer with a **cross peen** is useful for starting a small nail or tack. The peen is the tapering section where a claw would normally be. See Figure 4-21. When you turn the hammer upside-down, the peen is narrow enough to tap the head of the nail, striking it between your finger and thumb. Once the nail has sunk a little way into the wood, you can remove your hand, rotate the hammer, and finish the job.

In a cross-peen hammer, the peen is at right angles to the handle—that is, across it. In a **straight-peen** hammer, the peen is parallel with the handle; but this type is seldom seen.

A **sledge hammer** can have a head weighing 3lbs or more and can be used to drive stakes into the ground. It can also break up concrete, masonry, or just about anything else. A small sledge is shown in Figure 4-22.

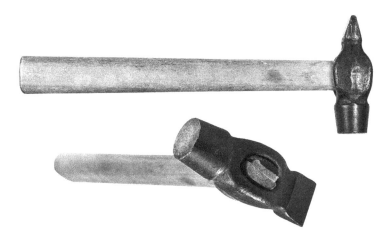

Figure 4-21. Two views of a cross-peen hammer.

A **rubber mallet** is useful for aligning delicate objects or nudging them into position, where the steel head of a hammer would cause damage. See Figure 4-23.

Hammer Substitutes

Various alternatives exist for inserting or removing fasteners. Here are just a few.

Figure 4-22. A small sledge hammer with a short wooden handle.

Brad Pusher

A **brad pusher** has a rounded handle attached to a thin magnetic plunger inside a hollow tube. If you are using small, thin nails known as brads (see page 51), you can load one into the tube, where the magnetic plunger will retain it. You force the brad it into wood by pushing the handle. This only works with soft woods. See Figure 4-24.

Figure 4-23. A well-used rubber mallet.

Figure 4-24. A brad pusher. Brads are described in the entry on brads in the section discussing nails.

Staple Gun

While electric versions of **staple guns** are available, the old hand-powered tool will work for you if you have a strong grip. This is much quicker than hammering small nails. If you look underneath an upholstered chair, you're likely to find that the fabric has been secured with staples. See Figure 4-25.

A special-purpose staple gun is available for round-topped staples that are ideal for securing low-voltage wires such as telephone or ethernet cables.

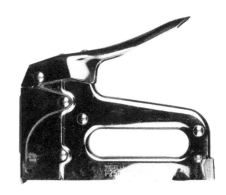

Figure 4-25. A manually operated staple gun. Some strength is needed to operate this device.

Riveter

Rivets are a quick way to fasten two pieces of thin material together. You drill a hole in each material, push a rivet through both holes, engage its tail in the riveter, and squeeze the handles. The underside of the rivet swells up, and then the tail snaps off.

Rivets are made of soft metal such as aluminum, and tend to be weaker than comparably sized screws or bolts. You have to use exactly the right length of rivet to match the thickness of the materials that you are fastening. This is a quick procedure, and you don't need access to the

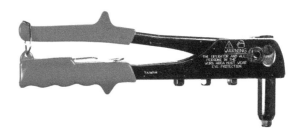

Figure 4-26. A basic manually powered riveter.

underside of the materials. A **riveter** is shown in Figure 4-26.

Nail Gun

A nail gun is loaded with special nails mounted in strips. Often powered by compressed air, the nail gun shoots a nail fully into a piece of wood when you press the trigger.

Heavy-duty power tools of the kind used on construction sites are mostly outside the scope of this book.

Pry Bar

A pry bar is designed for removing nails. See Figure 4-27. The sharply curved end functions like the head of a claw hammer, but spreads the force more widely, so it is less likely to dent or damage the wood. The other end of the pry bar can be inserted in a crack between two boards, or between a door and its frame, if you need to pry the pieces apart. The hole in this end of the pry bar allows it to be used on nails with flat heads that are sticking out. Pry bars are available in many sizes.

Figure 4-27. A pry bar.

Types of Nails

Figure 4-28, from left to right, shows a **brad**, a **roofing nail**, a **sinker**, a **ring-shank nail**, a **large finishing nail**, a **common nail**, a **drywall nail**, and a **tack**.

A **brad** is always thin and short, often used for attaching very small pieces of trim. It may be installed with a brad pusher, as shown in Figure 4-24.

Roofing nails have wider heads to spread the pressure across a greater area, so that the head is less likely to penetrate asphalt sheeting. They may be galvanized, or may be made of stainless steel to resist corrosion.

A **sinker** is coated with a thin film of adhesive that melts from friction when the nail is driven into wood. The adhesive cools almost immediately and then sticks to the wood, preventing the nail from coming loose. This type of nail is easily identified by the gold sheen of the adhesive. The head is embossed, to engage with a similar pattern in the striking face of a framing hammer.

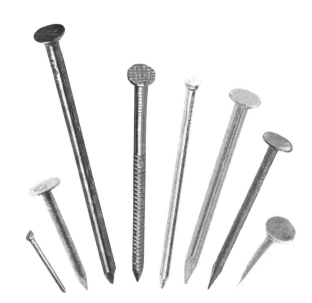

Figure 4-28. Some types of nails. See text for details.

Ring shank nails have ridges around the shaft (which is also known as the **shank**). The ridges help to hold the nail in place.

Finishing nails have small, rounded heads, the idea being that you can embed the head in the wood, recessing it so that it can be concealed with some caulking or wood filler, after which it can be painted over. To push the nail in below the surface of the wood, you can hammer it with a device designed for this purpose. A **nail set** has a flat or concave tip, while a **center punch** has a pointed tip, like an awl but heavier.

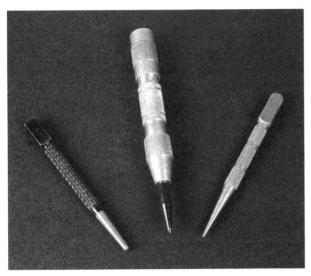

In Figure 4-29, nail sets are shown at left and right, while an automatic center punch is shown at center. A nail set must be hit with a hammer, while the automatic center punch triggers an internal spring-loaded mechanism when you push down on it hard. The center punch is appropriate for finishing nails, which have a small dimple in the head.

Drywall nails have a contoured head that is intended to recess itself in sheets of drywall that are being attached to wooden two-by-fours during the construction of a house.

Tacks taper to a very sharp point, and can be pushed into place with your thumb before being hit with a tack hammer. They are intended to secure fabric.

Figure 4-29. A nail set, a center punch, and another nail set.

If you wander around a big-box hardware store, you should find all of these and probably some other variants, too. Nails are usually not appropriate for small projects such as the ones in this book, but are still widely used in construction work and furniture upholstery.

Nail Sizes

In Europe, nail sizes are measured in millimeters, but in the United States, a very ancient system is used. A so-called "twopenny" nail is 1" in length, while a "threepenny" nail is 1¼" in length—and so on. The system is shown in Figure 4-30.

In England, before the money was decimalized, there used to be 240 pennies in a pound sterling, and a penny used to be represented with letter d. Why a d? Because it was an abbreviation for *denarius*, a Latin word that was used when the Romans occupied England a couple thousand years ago. When you buy a box of 1¼" three-penny nails with "3d" on the label, this is a piece of history from the very distant past. Look carefully, and you'll see the "3d" in Figure 4-4 on page 38.

Nails are sold by weight—typically in containers weighing 1lb, 5lbs, or (in some cases) 30lbs and even 50lbs. As the nails get longer, they also get fatter. But how many nails are in a pound? That depends. Figure 4-31 shows approximate quantities for common nails and finishing nails.

Penny Size	Length
2d	1"
3d	1 ¼"
4d	1 ½"
6d	2"
8d	2 ½"
10d	3
12d	3 ¼"
16d	3 ½"
20d	4"
30d	4 ½"
40d	5"
50d	5 ½"
60d	6"

Figure 4-30. A table of nail sizes. Sources disagree on the size of 20d and 30d nails, so I rounded the values to the nearest ½".

How a Nail Works

In a word: friction. When a hammer forces the nail into wood, the nail squeezes the wood, and the wood pushes back. This pressure prevents the nail from falling out. If the wood has not completed its drying process, it may shrink as it continues to dry, in which case the friction will diminish, and the nail may become less secure.

Screws create a much better connection with the wood, but they are more expensive, and driving a screw takes more time than hammering a nail. On a construction site, time is valuable, which is why nails still tend to be used.

I will explore all the attributes of screws in Chapter 10. Before I can do that, I'll have to deal with drilling holes, in Chapter 8. And before that, I'm going to explain how to make frames (such as picture frames), along with the wonderful attributes of plywood.

Penny Size	Common Nails per pound	Finishing Nails per pound
2d	850	1300
3d	550	800
4d	300	550
6d	150	300
8d	100	200
10d	70	120
12d	60	110
16d	50	90
20d	30	60

Figure 4-31. The number of nails you should expect to find in a 1lb box. Because sources disagree on the exact number, I rounded the values to the nearest 50 for nails up to 8d in size, and to the nearest 10 for larger sizes.

NEW TOPICS IN THIS CHAPTER

- Measuring angles
- Making mitered corners
- Non-rectangular frame shapes
- Making and using a jig

YOU WILL NEED

- One-by-six pine, no knots, not warped, length 12"
- Square dowel, ¼" x ¼", length 36"
- Square dowel, ¾" x ¾", length 60"
- Protractor (optional)
- Ratchet strap (optional)
- Utility knife

Also, as listed previously: Tenon saw, miter box, trigger clamps, ruler, speed square, rubber sanding block, work gloves, awl, masking tape, dust mask (optional), safety glasses (optional), plywood work surface, carpenter's glue, sandpaper, polyurethane, disposable gloves, paint brush (optional), finishing nails, and nylon rope or heavy string.

Check the Buying Guide on page 248 for information about buying these items.

If you look at any wooden picture frame, I'm betting you'll find that the corners are **mitered**. This means that the ends of each edge are cut diagonally so that they join neatly and symmetrically, as shown in Figure 5-1. The angle of the cut is 45 degrees, because that is halfway between 0 and 90 degrees. This is known as the **miter angle**. Your miter box is called a miter box because two of its slots are at 45-degree angles, to help you in making mitered corners.

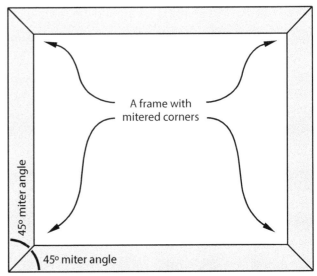

Figure 5-1. Almost all picture frames made from wood have corners mitered at 45 degrees. (Some plastic frames may be molded as a single piece.)

On the upside, mitering looks nice. On the downside, if you are using hand tools, making a mitered frame can be difficult.

But, not impossible! In this project, I'll show you how to make not only rectangular frames, but other shapes, too.

Your First Mitering Experience

The best way to learn is by practicing, so I suggest you use some of your square dowel to make a small frame measuring 6" x 5" along the outside edges.

In Figure 5-2, the first step is to draw two vertical lines 6" apart on your dowel. (Actual picture frames are usually sized by their inside dimensions, so that you know if your picture will fit, but for this project it's easier to start with the outside.)

Figure 5-2 Making the measurement.

Figure 5-3 Cutting a 45-degree angle.

In Figure 5-3, the dowel is in the miter box, and the saw is in the slots angled at 45 degrees. A cam could be used to hold the wood in place, but I assume you may not have one, as many boxes are supplied without them. You can use your thumb to hold the wood, but keep your fingers away from the saw blade. Cut carefully down the outside of the vertical pencil mark, while pressing with your thumb as hard as you can, as the saw will tend to push the wood around.

Check the length of your work, as in Figure 5-4. You may be tempted to smooth the fuzzy sawn edges with some sandpaper, but sanding will tend to spoil the accuracy of the cut.

Figure 5-4 Checking the length of the cut.

Now the good news: cutting your first piece at 45 degrees automatically creates a 45-degree angle on the remainder of the dowel, so that it can be used for next section of the frame. This is shown in Figure 5-5. Just make a new measurement, and cut that section, turn it around to fit the first, and continue until you have a total of four, as shown in Figure 5-6.

Figure 5-5. Cutting one frame piece automatically gives you the correct angle for the next frame piece.

Figure 5-6. Four pieces ready to be glued and clamped—somehow!

Check that the angles fit correctly. Now for the tricky part. How are you going to glue them together?

Clamping a Corner

Ideally you want to clamp each corner so that a diagonal force presses the mitered edges together, as shown in Figure 5-7. (I have exaggerated the amount of glue, to make it visible.) The problem is, if you try to use a clamp diagonally, it won't get a grip on the frame and will slip off.

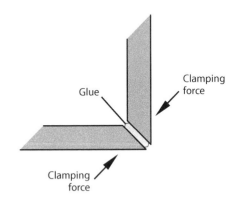

Figure 5-7. The problem of clamping a 45-degree corner.

This problem has been around for centuries, and many methods have been devised to deal with it. Just search Google Images for *clamped mitered corner*, and you'll see a great variety. But the fact that there are so many should tell you that none of them is absolutely ideal. If one method was better than all the others, no one would bother with the others anymore.

The simplest approach is to glue all the corners at once, run a strap around the outside of the frame, and tighten it. You can buy a strap specifically made for this purpose, but the cost is discouraging unless you are planning to frame a lot of pictures. The sort of ratchet strap that is sold as a tie-down for securing loads to pickup trucks or the roofs of SUVs will also work. You will want the type where the end feeds back into the ratchet. It should not terminate in a hook.

Figure 5-8 gives you the idea, and Figure 5-9 shows it working.

This works well for a small frame, but because the strap doesn't control the angles very well, a large frame may end up looking like Figure 5-10.

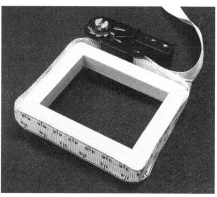

Figure 5-8. Using a ratchet strap instead of clamping.

Figure 5-9. A real-life ratchet strap in action.

Figure 5-10. Hazards of applying a ratchet strap to a large frame.

There are ways to prevent this—but what are the other options?

You may be wondering why we don't place two clamps across the frame at 90 degrees to each other, as shown in Figure 5-11. There are two problems with this arrangement. First, if you have a large frame, the clamps will tend to bend it, which will cause the mitered corners to open a little. You can overcome that by inserting some two-by-fours between the clamps and the frame sections, but the second problem is that you have to get the pressure and spacing of each of the clamps exactly right—otherwise, you end up with a situation shown in Figure 5-12.

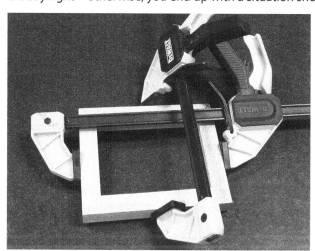

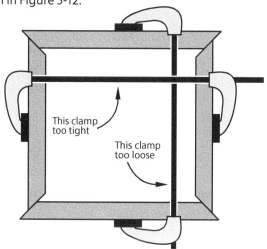

This clamp too tight

This clamp too loose

Figure 5-11. A simple option is to put clamps across the frame at 90 degrees to each other.

Figure 5-12. This will tend to happen if the clamps are not set precisely.

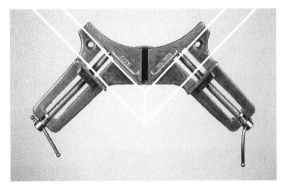

Figure 5-13. A ready-made corner clamp.

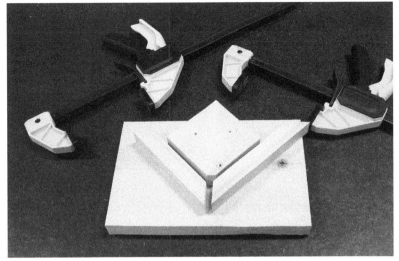

Figure 5-14. A simple jig to make a 90-degree frame corner.

Figure 5-15. The frame pieces clamped into position.

Another option is to buy corner clamps, such as the one in Figure 5-13. I superimposed a white outline to show how the frame fits into this type of clamp. But to deal with all four corners simultaneously (which is the best way), you'll need four clamps, and they are not cheap.

I think a good answer is to make a **jig**.

Making and Using a Jig

A jig is an Irish dance, but in a workshop, the word describes something that you build, to help you to align the parts in a project that you are working on. (The jig may have acquired its name because it stops parts from jiggling around.)

I try to avoid using jigs, because the time that I spend making them seems unproductive. After all, the jig has no use after the real project is done. However, a mitering jig is very simple and actually can be reused—assuming you put it in a safe place, and then remember where you put it.

My design works very much like a corner clamp, and is shown in Figure 5-14. A piece of pine, ¾" thick and 3" x 3" square, with a corner cut off, is nailed to a piece of one-by-six pine, using the same 1¼" finishing nails that were used in Chapter 4. Sections of the frame will be clamped to the square, as shown in Figure 5-15.

Making the jig is very easy. First cut 12" from the end of a one-by-six board, using the same method as in Chapter 4 (see Figure 4-9 on page 41). Now you need to remove 3" from the end of the

12" piece, again using the same method as in Figure 4-9. Set aside the remainder, which will become the base of your jig, as shown in Figure 5-15.

Turn around the piece that you just cut, measuring 3" x 5½", and draw a line 3" from its end. Put it in your miter box, and cut along the line, being careful that the cut is very accurate, as it will be controlling the accuracy of your frame corners. If the cut doesn't look good, try cutting another ¼" off it, as the specific size of the block that you are making is not crucial. Remember:

A project that you build with a jig is only as accurate as the jig itself.

A plan in Figure 5-16 illustrates the three saw cuts that I just described.

One corner of the 3" x 3" square has to be beveled off, to keep it out of the way if glue squeezes out of the joint in the frame. You can bevel the corner by trimming it with a saw, or by sanding it. The precise amount is not important.

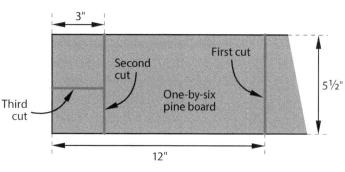

Figure 5-16 Cutting a 3" x 3" square from a one-by-six board.

Position the 3" x 3" square as shown in Figure 5-15, hammer in a couple of nails, and your jig is ready for use. The base is not strictly necessary, but provides stability and makes the jig easier to handle.

You may wonder how the clamps apply force to the miter joint. Each clamp is exerting pressure between the frame and the center square, but not to the joint itself. This is true, but after you apply glue to the joint and position everything tightly, there is just enough flexibility in the setup so that the pressure of the clamps will push it all together, including the joint.

The only problem with this system is that if you work on one corner at a time, and you make even a small error as you add each section of the frame, the last section may not fit precisely with the first section. However, if necessary, you can sand or trim the last sections to fit.

Is this the best way to miter a frame? No, I never made that claim. My design is just simple and cheap. Also, I wanted to mention the concept of a jig, because it's important, and will be necessary in chapters 9 and 19.

The "best" way to miter a frame is probably to spend hundreds of dollars on a gadget designed for that one purpose. But I assume you would prefer not to do that. And in any case, the gadget would be of no use if you ever want to make a non-rectangular frame—which is what I want to do now.

Non-Rectangular Frames

Take a look at the frames in Figure 5-17. Is it easy to make shapes like these? I think it is. All you have to do is transfer the miter angle, which I have shown, to pieces of wood. And I have figured out three different ways to do that.

First, let's give these things some names. The proper name for a shape that has multiple straight sides is a *polygon*. If the sides are all of the same length, and all have the same angles to each other, you have a *regular polygon*.

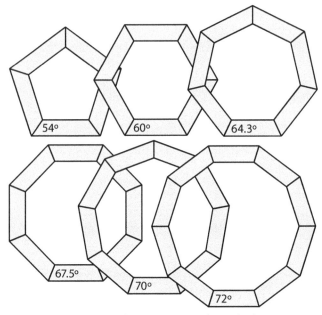

Figure 5-17. Frames with interesting numbers of sides.

The names for the polygons in Figure 5-17 are:

- 5 sides: pentagon
- 6 sides: hexagon
- 7 sides: heptagon
- 8 sides: octagon
- 9 sides: nonagon
- 10 sides: decagon

Now that I've shared that bit of trivia, I'll explain the miter angles. One way to draw angles is to buy an ancient device known as

Figure 5-18. A protractor.

a protractor, shown in Figure 5-18. Protractors are still available in stationery stores. Draw a straight line on a piece of paper and make a mark on the line. I'm going to call that mark the *origin*. Align the base of the protractor with your line, and put its center mark over the origin. Now make a mark beside the number of degrees that you want. Remove the protractor, draw a line to the mark from the origin, use scissors or a knife to cut along the lines that you drew, and transfer the angle to the wood by drawing around the paper. When you use a paper shape like this, it is called a **template**.

What if you don't want to buy a protractor? You can search for an image of one online which can be printed on paper—although this won't be as easy to use as transparent plastic. Alternatively, perhaps you have some drawing software, and you can tell it to give you any angle you want on your computer screen. Print it, and once again, you have a template.

What if you don't have drawing software? No problem! I wrote a little computer program that found some dimensions of triangles that just happen to have the necessary miter angles in them. Take a look at Figure 5-19.

This shows you that if you start with a rectangular piece of paper, and you measure a distance along the vertical edge named V, 128mm from the bottom-right corner, and a distance named H which is 93mm horizontally from the bottom-right corner, and connect the ends of the lines, the angle at the bottom left corner just happens to be almost exactly 54 degrees, which is the miter angle for a five-sided frame.

You can cut the triangle out of the piece of paper, apply it to the wood you are using for your frame, and draw along the diagonal line to make your miter angle. See Figure 5-21. Then you will be very, very careful to saw along the line freehand, because your miter box cannot be used for this job. It does not have a slot for 54 degrees.

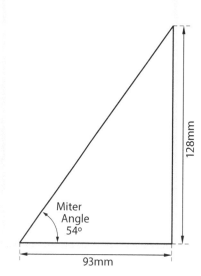

Figure 5-19. Help in creating a miter angle.

The table in Figure 5-20 tells you values for V and H that will give you other miter angles. The angles will not be absolutely, totally precise, but they will be more accurate than is possible when working in the real world with hand tools. So, they are good enough.

If you are wondering why I am using millimeters for these measurements, it's for the same reason as in Chapter 4. They make measurements on paper easier than if I was asking you to use 32nds or 64ths of an inch.

What if you want to make a polygon with more than 10 sides? You'd have to figure out the miter angle yourself. Just in case you want to know, I'll tell you how.

Number of Sides	Miter Angle	Distance V (millimeters)	Distance H (millimeters)
3	30°	56	97
4	45°	100	100
5	54°	128	93
6	60°	168	97
7	64.3°	164	79
8	67.5°	239	99
9	70°	272	99
10	72°	317	103

Figure 5-20. Table of miter angles, with V and H values in millimeters for drawing them on paper.

If your polygon has N sides, the miter angle = 90 − (180 / N).

For instance, if N = 12, the miter angle = 90 − (180 / 12).

This would be 90 − 15 = 75 degrees.

That's all there is to it.

Making Your Own Pentagon

To test this system, I suggest you cut a frame in the shape of a pentagon, which I have always thought is a very nice-looking polygon (even though it is also the shape

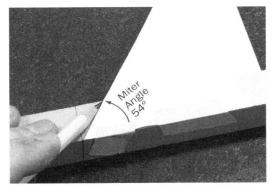

Figure 5-21. Getting ready to cut miter joints for a pentagon-shaped frame.

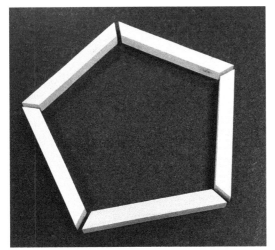

Figure 5-22. Five frame pieces.

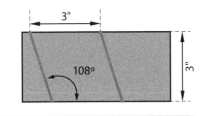

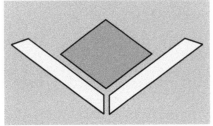

Figure 5-23. A jig for a pentagon.

of a rather large, ugly building in Washington D.C.). In Figure 5-21 I have marked 7 inches on a piece of square dowel, because I decided that this will be the external length of each side of the frame. I am using the template to show where to make my cuts.

Figure 5-22 shows the five frame pieces. How can you glue them together? You can make a jig, as I described before, except it won't have a square block. You would have to make the block using the plan in Figure 5-23

The upper half of the plan in Figure 5-23 shows how to cut the jig out of a 3" piece of ¾" pine. You can create the 108-degree angles using a protractor, or using drawing software, or by putting two 54-degree templates together. Both of the red cut lines have to be at the same angle, because clamps will only work reliably if they are applied to parallel edges.

The lower half of the plan shows how the frame edges would be grouped around the jig.

If that seems like too much trouble, this is a small enough frame for a ratchet strap to work—if you have one. If you don't, you could use a piece of thick nylon rope with a slip knot, and pull it as tightly as possible around the frame. After you pull the knot tight, you can stop it from loosening by sticking a nail or a pick through it. Avoid using thin rope or string, unless you slip in some cardboard to stop the rope from digging into the wood. See Figure 5-24. After allowing time for the glue to set, release the frame and sand it to compensate for any slight unevenness in the joints.

Actual Picture Frames

I've been using square dowels for this project because you have dealt with them previously. An actual picture frame would have a channel cut into it, to hold a picture.

You can buy frame sections at a crafts store, but there is an alternative. You can add a thin strip of trim around the interior of your dowelled frame.

First trace the internal shape of your pentagon frame onto a piece of paper. You will need to tape two sheets of paper together, or use a large sheet, as in Figure 5-25.

Now you need some trim that is similar to a square dowel but only ¼" x ¼" when viewed from the end. You can find this trim cheaply in a crafts store. Line it up with the inside of the pentagon that you drew on paper, and make miter cuts in the trim.

It's so thin and delicate, I suggest you cut it with a utility knife, as in Figure 5-26. Just push down hard while wiggling the blade a little. Cut it slightly larger than it should be, and then sand it to fit.

Now you can glue it around the inside edge of your pentagon, as in Figure 5-27. You can hold it in place using tape, or cable ties, or anything else that applies moderate pressure, until the glue sets.

Do you want glass in your frame? I'm not so sure about that. Glass is difficult to work with, and the edges are wickedly sharp.

You could use **polycarbonate** instead, often sold under the trade name **Lexan**. Just visit your local hardware store, and you can find it in small sheets, intended for replacing broken window panes. You can cut it with a saw, but I'll come back to this when I discuss transparent plastic in Chapter 18. You may want to put the frame aside till then.

Figure 5-24. A pentagonal frame, roped while the glue sets.

Figure 5-25. Tracing the pentagonal frame onto paper.

Figure 5-26. Cutting the trim to fit.

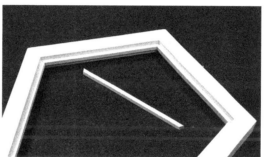

Figure 5-27. Pentagon trimmed out.

Figure 5-28. The shape of a subject should always match the shape of its frame.

The trim in the frame will retain your polycarbonate panel and a photograph behind it. You can add a piece of cardboard at the back, which will push-fit or can be held in place with very small brads or pins. Just bear in mind that when you choose a picture to put in your frame, it should be appropriate to the shape of the frame. An example is shown in Figure 5-28.

Ubiquitous Geometry

This project has been all about angles. You may feel that this is a slightly obscure topic, but understanding angles can be useful. For instance, if you ever want to frame the peaked roof of a house or the slanting roof of a shed, angles are unavoidable. Your speed square has angles marked on it for this purpose.

As for polygons—how about if you want to make an eight-sided dining table? Or a five-sided coaster? Or a six-sided jewelry box?

Figure 5-29. A 12-sided tower under construction.

I once designed a 12-sided tower to stand on top of a small mountain. The tower under construction is shown in Figure 5-29. We had to bevel the framing to fit. This required me to know the miter angle, which was 75 degrees. Do you remember how to calculate that?

Geometry is all around us. Some pots to hold pens and pencils are six-sided. You may see eight-sided windows in houses. The homes in some Native American tribes were eight-sided.

Symbols are often created with polygons. In some counties in the United States, a sheriff's badge has seven points. A stop sign is eight-sided. A five-pointed star is a common symbol, not just in the United States but also Russia and China. The easiest way to draw that star it is to start with a pentagon.

Filling Gaps

If you make small errors in cutting frame sections, gaps will occur. What to do about this? Well, how about filling the gap with a wood filler such as Plastic Wood? This can be used by people who don't always manage to make things fit precisely (people such as me, for instance).

Plastic Wood is sold in little cans, in a variety of colors. The colors match the wood that you are working with, so in theory, no one will notice it. Be warned, however, that if you apply polyurethane, Plastic Wood will turn a completely different color from the real wood beside it, and it won't be invisible anymore.

Figure 5-30 shows an example of this problem. The filler was totally invisible, until after it was polyurethaned, at which point—well, you can see for yourself.

Also be warned that Plastic Wood actually sets harder than oak. You will have difficulty removing any smears that shouldn't be there. Other brands of wood filler are available that don't set so hard, but I haven't done an exhaustive comparison test.

Add it up, and your life will be easier if you can work without wood filler. One way around it is to sand your work, allowing fine sawdust to accumulate in the gaps. Wipe the dust away from the flat surfaces, but leave the dust in the gaps, and apply a quick coat of polyurethane.

Other Ideas

A Heart-Shaped Frame

Take a look at the design in Figure 5-31. Do you think you could make that? Cutting the angles would be easy, because all of them are either 45 degrees or 67.5 degrees, and those just happen to be the angles in your miter box. (In some miter boxes, the 67.5-degree miter angle is described as 22.5 degrees, for reasons illustrated in Figure 5-32.)

Clamping this frame would be a challenge, but you can make a jig as suggested in Figure 5-33.

Figure 5-30. What happens to plastic wood labeled "red oak" in color, when it is applied to red oak and then coated in polyurethane.

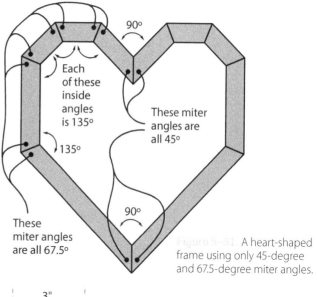

Figure 5-31. A heart-shaped frame using only 45-degree and 67.5-degree miter angles.

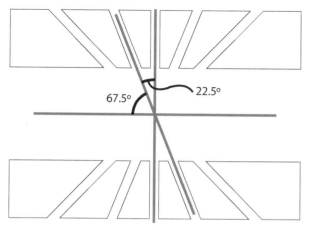

Figure 5-32. Overview of a miter box showing the 67.5-degree slot, which is sometimes referred to as being 22.5 degrees.

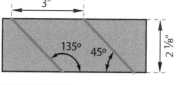

Figure 5-33. Making a jig for the angles labeled as 135 degrees in Figure 5-31.

Clamp Fact Sheet

To minimize your costs, I have been making all the projects in this book using just two trigger clamps. But of course, many other clamps are available.

Basic Screw Clamps

These can range from tiny to massive, depending on the job. In The US, they are called C clamps, while in the UK they are called G clamps. Personally I think they look more like a G than a C. A medium-sized example is shown in Figure 5-34.

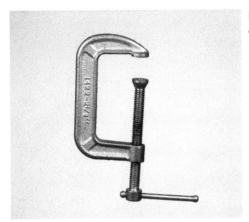

Figure 5-34. A basic C clamp.

Quick-Release Screw Clamps

Shown in Figure 5-35. These are like trigger clamps, except that after you slide the clamp into its approximate position, you apply the final amount of force by rotating a handle attached to a screw thread. This is much more powerful than a trigger clamp.

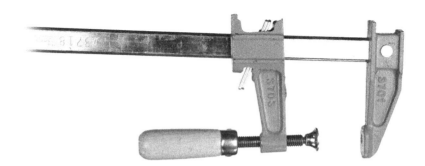

Figure 5-35. A quick-release screw clamp.

Pipe Clamps

Useful when you need to clamp very large items, such as pieces of furniture. A sample is shown in Figure 5-36. They are not cheap!

Figure 5-36. A four-foot pipe clamp.

Locking Clamps

Vise-Grip clamps or pliers are a brand name that is often used for any locking clamp that maintains its hold till you release it with an unlocking lever. Dozens of variants are available. A locking C clamp is shown in Figure 5-37.

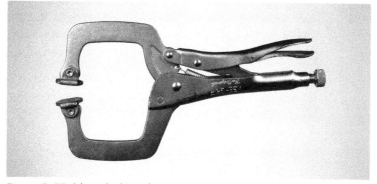

Figure 5-37. A large locking clamp.

Butt-Joint Clamp

Can be used to hold two piece of wood together at 90 degrees. A vertical section is inserted in the upper part of the clamp, and a horizontal section is inserted below it. The lower threaded clamp pushes the horizontal section up flush against the bottom of the vertical section. See Figure 5-38.

Figure 5-38. Butt-joint clamp.

Chapter 6
A Geometrical Jigsaw

YOU WILL NEED

- Two-by-four pine, any condition, length 12"
- Square dowel, ¾" x ¾", length 30"
- Plywood, pine or oak, ¼" thick, size 12" x 18"
- Wood stain, dark, minimum quantity

Also, as listed previously: Tenon saw, miter box, trigger clamps, ruler, speed square, rubber sanding block, work gloves, awl, masking tape, dust mask (optional), safety glasses (optional), utility saw (optional), plywood work surface, carpenter's glue, sandpaper, polyurethane, disposable gloves, paint brush (optional), and nylon rope or heavy string.

Check the Buying Guide on page 248 for information about buying these items.

Making a jigsaw puzzle that has interlocking pieces is too ambitious for this book, but other types of jigsaw are possible. This project shows you how to build one that looks simple but is actually more difficult to assemble than the traditional type. And it's a great way to become familiar with a wonderfully versatile material: **plywood**.

Where Plywood Comes From

When layers of wood are combined, we say that they are **laminated**. The concept of lamination stretches back even to the Egyptians, but plywood as we know it today is about 100 years old. It is manufactured now as it was then, by peeling layers of wood from a rotating log. Multiple layers are then glued together to create sheets.

Plywood is one of many wooden **composites**, meaning materials that are composed of small pieces or fragments, usually glued together. For additional details, see the Composites Fact Sheet on page 78.

The Strength of Plies

To see why plywood is so useful, I'd like you to do a little test. Start with a piece of scrap two-by-four, and cut a piece about ¾" long from the end of it. (The exact amount is not important.) Turn it so that the grain is facing upward, and clamp it to your work area as shown in Figure 6-1.

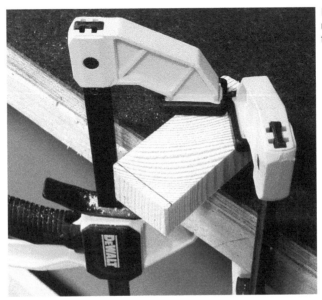

Figure 6-1. Preparing to remove a thin slice from the end of a two-by-four.

Saw a slice from this piece, across the grain, about ⅛" thick, as shown in Figure 6-2. Because this is a small slice, be especially careful about the saw jumping out. Keep your free hand away from the cut.

Take the slice between your fingers, as shown in Figure 6-3, and apply some bending pressure. You'll find that the wood easily snaps in your fingers, as in Figure 6-4.

Now take one of the broken pieces and try to snap it in the other direction, along a line *across* the grain. You'll have to use much more force, and you may have trouble breaking it at all.

The weakness of wood when it is stressed parallel with the grain can be a problem when structural strength is an issue. Furniture, for instance, has to be designed with this in mind. In addition, joining boards to make a large flat area is a time-consuming chore.

Plywood eliminates both of these problems. It is usually sold in sheets 48" x 96" (sometimes larger), and its layers, or **plies**, are glued at right angles to each other. Over any fixed distance, plywood should be equally strong in both directions.

Plywood is also relatively stable, meaning that it doesn't tend to warp like a board or a beam. It comes in many different grades and thicknesses.

Figure 6-5 shows five samples. From top to bottom, they measure ⅛", ⁷⁄₃₂", ¼", ⁵⁄₁₆", and ⅜" in thickness. The three pale-colored pieces are high-quality material from a crafts store. The piece second from the top came from a big-box hardware store and was their best-quality material in that thickness.

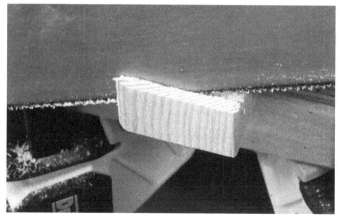

Figure 6-2. Cutting the slice across the grain.

Figure 6-3. Bending along a line parallel with the grain.

Figure 6-4. The wood is easily broken parallel with the grain.

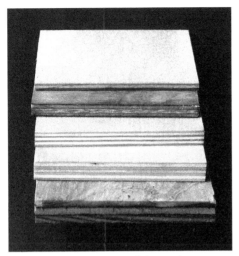

Figure 6-5. Five samples of plywood. See text for details.

Figure 6-6. High-quality ¾" plywood.

The piece at the bottom, measuring ⅜" thick, is the cheapest of the bunch, with only three plies and a rough surface of soft wood.

In Figure 6-6 you see a piece of high-quality ¾" plywood with nine plies and a birch surface. Again, this came from a crafts store. The type of ¾" plywood used in construction work usually has a rough finish and cannot be used for work on a small scale where the quality and detail of the finish are important.

I chose the ⁷⁄₃₂" plywood for the geometrical jigsaw. If you try to work with material that is thinner than this, you'll have problems with it splintering. If you opt for thicker material, you'll do more work sawing it. I think plywood that is no more than ¼" thick with a hardwood surface is an appropriate choice.

The Pattern

My plan for the jigsaw is shown in Figure 6-7. It looks simple, but you'll find it is cunningly designed so that assembling the pieces is a challenge. Some of them have equal edge lengths, allowing them to fit together in many different ways, while others have edges that are not quite equal, to create confusion. I have also included a lot of 90-degree angles, so that it's difficult to tell which pieces go in the corners.

I can think of three ways to transfer this pattern onto paper, so that you can then apply it to a sheet of wood. The simplest option is to photocopy the page from the printed version of this book, or copy-paste the image if you are viewing it on a computer screen. You can enlarge the image when you print or copy it, but it should not exceed 6" in the longest dimension, because you'll have trouble making a longer cut with your tenon saw—as you will soon see.

Another option is to draw the pattern yourself, on graph paper. Stationery stores sell graph paper, or there are many web sites which allow you to download it free, after which you can print it.

A third option is to reproduce the pattern on a computer using vector-graphics software. I'm going to discuss that in detail, because this kind software is useful if you want to create designs of your own. And the application that I am recommending is free.

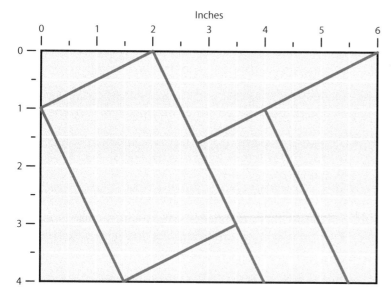

Inches

Figure 6-7
The geometrical jigsaw.

Vector-Graphics Software

If you are using photo-editing software, and you enlarge the image, the pixels get bigger—unless you resample the image, in which case you lose detail. This kind of software is not appropriate for accurate line drawings.

In **vector-graphics** software, the image is saved as a set of numbers defining lines and curves. A rectangle of a specified size will be reproduced at exactly that size, because the numbers store that information. If you double the size, you won't lose detail, because the software simply doubles all the numbers that describe the shape.

Many vector-graphics programs exist, but the one that I suggest is OpenOffice Draw, because it is freeware for Windows, Mac, and Linux.

Go to www.openoffice.org and click the Download tab. Be careful that you don't visit a site that has a name similar to openoffice; there have been some spoof sites that add unwanted toolbars or search bars to your download.

You will have to download the whole OpenOffice suite of applications, but after that, if you wish, you can install just Draw.

When you complete the download, start the installer and answer the usual questions until you reach the option to do a *typical* install or a *custom* install. Choose *custom*.

Now you can select the Draw component, and you can deselect everything else— unless, of course, you want the other applications. Writer, for example, emulates Microsoft Word, while Calc emulates Microsoft Excel. These applications can read Microsoft files, and can save in Microsoft formats.

Using OpenOffice Draw

When you launch OpenOffice, you'll be able to select the Drawing feature, and it will show you a blank document. Here are the basic functions that you need to know.

Begin by selecting View>Grid from the menu bar, and make sure that Display Grid and Snap to Grid are both selected (click the little icon beside the option for each one, to toggle it on and off).

You should now see a grid of dots on the screen. They are pale gray and not easily visible, but we can fix that in a moment.

Select Tools>Options from the menu bar, and in the dialog box that opens, double-click OpenOffice at the top of the list, then Appearance in the submenu that appears. Scroll down to Drawing/Presentation and you can change the grid color (although it won't become visible yet).

Go back to the mini-menu on the left side of the dialog box, double-click General, and you can choose your units of measurement. (On some platforms, this option is under the OpenOffice Draw menu.) I'm assuming you will want inches, but you can select millimeters if you wish. Now double-click Grid in the submenu, and under Resolution, you can choose the horizontal and vertical spacing of the bold dots in the grid. You can also specify the number of subdivisions. The default may be 5 or 10 (again, it depends which version and which platform you are using), but either way, I suggest 16, to make your drawing compatible with your stainless-steel ruler. Click OK, and the page on your screen should change in accordance with your settings.

Now you can use the line tool, the rectangle tool, or the ellipse tool to draw shapes on the screen. These tools are located at the bottom-left corner. Click on any tool to activate it.

You create each shape by dragging the mouse. Hold down Shift to draw lines at preset angle increments, or to create a square instead of a rectangle, or a circle instead of an ellipse.

After you draw an object, it selects itself, with little handles that pop into view. Now you can change the line thickness, line color, and fill color using data-entry fields at the top-left of the screen. Also, if you right-click an object (option-click, on a Mac), you get a menu which allows you to modify the object in various ways. Select Position and Size, and you will find options to resize, move, or rotate the object by exact amounts.

The crucial thing to remember, when using this software, is that you must have one or more objects selected before you can do anything with them. The Arrow tool (at bottom left, on the screen) allows you to click on any object, or select multiple objects by stretching a marquee (dashed lines) around them.

The program defaults back to the Arrow tool after you create any new object.

Draw is not a very elegant program, and it lacks many of the features that you find in professional-level software such as Adobe Illustrator. We shouldn't be too critical of it, though, because it will do as much as you need for the projects in this book, and it costs nothing at all.

Awl You Need

Assuming you have printed the pattern onto paper one way or another, your next step is to transfer it onto your plywood. You can't just tape the paper to the wood and start sawing, because the saw will rip and mash the paper. You'll have to use an awl.

Just prick through each corner of the pattern into the wood, as in Figure 6-8, then remove the paper and draw lines between the prick marks, as in Figure 6-9.

Cutting a Work Piece

Assuming that your plywood is substantially larger than the puzzle, you should cut roughly around the puzzle to obtain a work piece that will be easier to deal with. Leave a margin of about an inch.

This is a rough cut, so you don't have to worry about its exact position or quality. In fact, you would have difficulty putting sacrificial wood under the plywood, because the saw would have to cut through it. See Figure 6-10.

The rib along the top of the tenon saw makes it less than ideal for this long saw cut, but it can do the job.

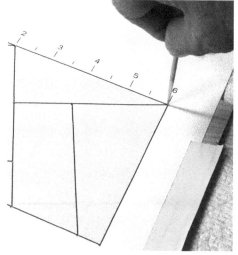

Figure 6-8: Tape the paper pattern to your plywood, and prick through each corner with an awl or a pick.

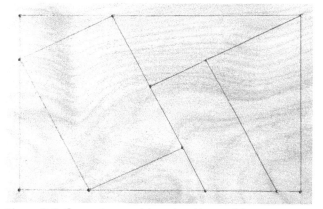

Figure 6-9: Connect your prick marks with pencil lines.

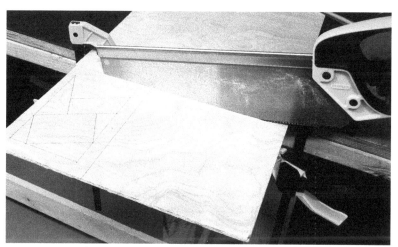

Figure 6-10: Making a rough cut to obtain a piece of plywood that will be easier to work with.

Figure 6-11. The underside of a saw cut when a sacrificial piece of wood is not used to reduce splintering.

Figure 6-12. Staining the back side of your plywood.

In Figure 6-11, you see what happens to the underside of plywood when you don't bother to reduce splintering. In this case, it doesn't matter, as this edge won't be used in the puzzle pieces.

Start With Some Stain

Thinking ahead, I want the top surface of the jigsaw to look different from the bottom surface, because if someone doesn't know which way up the pieces are supposed to be, the puzzle will become almost impossible to solve.

The easiest way to make the underside look different is by staining it. Wood stain is easy to apply and dries very fast, but there are some aspects you should bear in mind.

▪ Stain will be difficult to remove from hands, clothing, and work surfaces. Wear disposable gloves, dress appropriately, and protect the area where you are working.

▪ Shake the can vigorously before opening it, and use a stirrer from time to time while the can is open, to make sure that sludge doesn't accumulate at the bottom of the can.

▪ Some types of stain are available in oil-based and water-based versions. I prefer the oil-based, but it must be used with good ventilation. Also, it is a fire hazard; you should not accumulate rags or paper towels with oil-based stain on them inside your house.

▪ The color of the stain will be affected by the color of the wood that you use it on.

▪ Check the label on the can carefully for additional cautions.

In Figure 6-12 you see how a paper towel can rub stain over the wood. Stain penetrates into the grain, and the longer you leave it, the darker it will get. After allowing it to soak in, you should remove any surplus from the surface by rubbing it with a clean piece of towel. Ten minutes later, more or less, you may be able to handle the wood without getting stain on your fingers.

The Cut Sequence

Now you're ready to cut the actual pieces of the jigsaw. Does it matter which pieces you cut first? Absolutely! In Figure 6-13, I have lettered the cuts in alphabetical sequence. The rough edges indicate the rough cuts that you made to extract this piece from your larger plywood sheet.

The reason that the sequence is important is that each cut must go all the way from edge to edge of the piece you are cutting. Where you have a T-shaped intersection of cuts, the sacrificial piece that you are going to put underneath the plywood makes it virtually impossible to cut the vertical stroke of a T first. You have to begin with the stroke from edge to edge. Then you can do the vertical stroke, from edge to edge of the smaller piece. Figure 6-14 illustrates this.

Accurate Cutting

Accuracy is important in this project, because the pieces of the jigsaw have to fit together. Therefore, you will need a guide piece on top of the plywood, as well as a sacrificial piece below it. (I described these concepts back in Chapter 3. See page 27.)

Figure 6-15 shows the setup, and Figure 6-16 shows the first accurate cut in progress, along the edge of the jigsaw

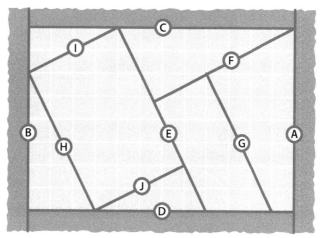

Figure 6-13. The sequence for cutting the puzzle pieces.

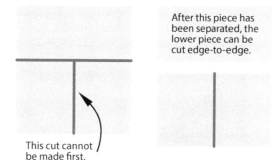

After this piece has been separated, the lower piece can be cut edge-to-edge.

This cut cannot be made first.

Figure 6-14. Each cut must extend from edge to edge of the piece that you are currently working on.

Figure 6-15. The plywood is clamped between a guide piece of two-by-four, on top, and a sacrificial piece, underneath.

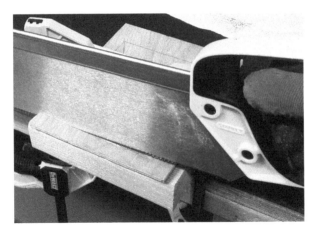

Figure 6-16. Cutting the edge of the jigsaw rectangle.

Figure 6-17. Each cut separates pieces of the puzzle, and therefore the saw must cut along each line, not beside each line.

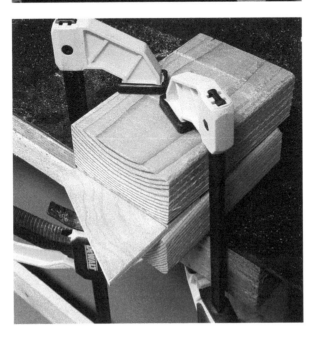

Figure 6-18. Ready for the last few cuts.

rectangle. Note that in this project you must cut *on* the lines, not *beside* the lines. This is because many of the cuts pass between the puzzle pieces, and you want to remove an equal amount of wood from each piece.

The saw has to tilt at a slight angle, because if you try to use it horizontally, it will tend to skate around instead of digging into the wood. The angle of the saw inevitably means that it will cut a groove in the sacrificial piece. That's what the sacrificial piece is for, but it does mean a bit of extra work for you.

In Figure 6-17, the job has progressed to the cut marked E in Figure 6-13. In Figure 6-18, the last few cuts are being made, and are short enough to allow your guide piece and sacrificial piece to be turned around.

Finishing

Figure 6-19 shows the finished puzzle pieces. But wouldn't it be better if you enclosed them in a frame, as you did with the sliding block puzzle in Chapter 1? Since you already learned how to make a mitered frame, this shouldn't take long. The frame I built is in Figure 6-20. I added a back to it, using the remainder of the plywood.

If you ask someone to fit the pieces into the frame, I guarantee they'll have a hard time—unless they realize that they can match the grain on the pieces, like the picture on a conventional jigsaw. Maybe you should prevent this?

The obvious way would be to cover up the grain with paint. On the other hand, maybe you would prefer to leave the grain visible, to provide a clue for solving the puzzle—in which case, you could polyurethane the pieces. I left my version unfinished. You can choose how to finish yours.

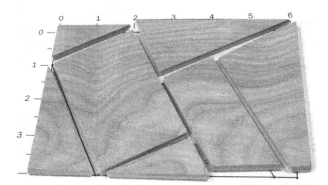

Figure 6-19.
The finished pieces.

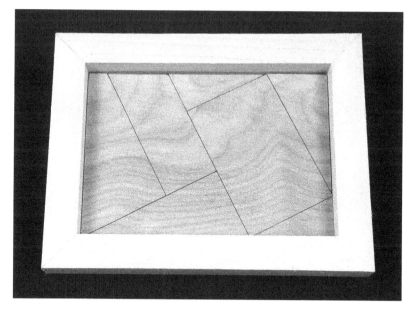

Figure 6-20.
The puzzle, framed.

A wooden composite is generally a sheet that is composed of thinner or smaller pieces of wood, glued together. The most common composites are **plywood**, **particle board**, **MDF**, **Masonite**, and **OSB**.

Plywood is made in grades identified by letters A, B, C, and D, the A grade showing the fewest defects while the D grade may have knot holes and other major flaws. D-grade plywood is only suitable for rough construction work.

Some plywoods are equally good on both sides, in which case they can receive a grade such as A-A or B-B. When the sides are not of equal quality, the wood may receive a grade such as A-B or A-C.

The cheapest plywood contains only three layers of softwood, and is sometimes referred to as **three-ply**. Obvious grain patterns will be visible on the surface, in the same way that grain patterns are visible on a board of soft wood.

Higher-quality plywood will have more, denser layers and may be faced with hardwood such as oak. The edges will be much cleaner and smoother than the edges of three-ply.

This is colloquially known as **chipboard**. It is made of compressed sawdust, glued together. Because the surface does not look very appealing, it is almost always covered with a wood veneer, vinyl, or a laminate such as **Formica**.

Because particle board contains no grain of any kind, it does not usually warp, and is a good choice for kitchen cabinets and cheap furniture such as bookcases. It bends and breaks much more easily than plywood of equal thickness, and its granular structure tends to disintegrate when screws or nails are driven into it. Shelves made from particle board are often joined to vertical supports using wooden pegs in holes, or very long screws with ample-diameter pilot holes.

This product is vulnerable to moisture. When it soaks up water, it swells and loses structural strength.

This resembles particle board, but is made from fibers rather than sawdust, costs a little more, and is less liable to chip and break. Its name is an acronym for **medium density fiberboard**.

MDF may be found in loudspeaker cabinets, where its high density lowers the resonant frequency.

Low-cost closet shelving can be bought from big-box hardware stores and lumber yards, prefinished with a thin white layer that is similar to Formica but thinner and less robust. This layer is known as **melamine**, and may be preapplied either to particle board or MDF. Confusingly, people often refer to the combination of the finish and the board as melamine.

In some precut products, at least one edge is finished in the same way as the surface of the board. The shelving can be trimmed to size by sawing the ends and the back edge, and can then be installed without any need for painting.

This was developed in the 1920s and named after its inventor, William Mason, who trademarked the name of the product. It is made by saturating wood chips in steam at a very high pressure. This reduces the wood to a mass of fibers that are compressed to form dark brown sheets with a distinctive smooth finish on one side and rough, textured surface on the other. No glue is used in this process.

Typically, Masonite is sold in thicknesses of ⅛" and ¼", suitable for very cheap paneling in applications such as closet doors that have a hollow interior. Masonite that is prepainted white on one side is sold as **bathroom board** (sometimes, **showerboard**), and may be embossed with fake tile patterns. It is the cheapest of all composites.

Hardboard is a generic version of Masonite.

This three-letter acronym is derived from **oriented strand board**, containing strands of wood in successive layers oriented at 90 degrees to each other. The low cost of OSB has enabled it to displace plywood in applications such as the floors of houses built in the United States. OSB has a rough surface and very rough edges, and is not suitable for higher-quality purposes such as furniture-making.

Regardless of which kind of composite you're using, the problem of cutting it remains the same. The stiffening rib along the top of a tenon saw won't pass through the cut that you're making. If the cut is a very long one, a tenon saw becomes difficult to use, and impossible if you want to include a sacrificial piece to minimize splintering.

Hand-sawing plywood really requires a panel saw (which you can push all the way through the cut) or a Japanese-style pull-saw (which you can pull all the way through the cut).

The easiest way of all is to use a circular saw, but as I've mentioned before, I think you need an experienced person who can be with you during your first adventures with power saws.

NEW TOPICS IN THIS CHAPTER

⬛ Parquetry
⬛ Using epoxy glue

See next page for a materials list.

Marquetry is the fine art of creating pictures by fitting together **veneers** (thin layers of wood), often to decorate furniture. An example is shown in Figure 7-1, more than 250 years old.

Parquetry is similar to marquetry, but uses simpler, abstract patterns of repeating shapes. **Parquet flooring** is a form of parquetry, as its name suggests. Parquetry can be as simple as the traditional patterns in Figure 7-2, Figure 7-3, and Figure 7-4, or as complex as the amazing example in Figure 7-5.

Figure 7-1. A fine piece of marquetry displayed at the Chazen Museum of Art. From Wikimedia Commons.

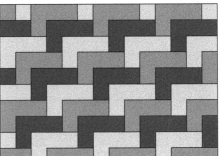

Figure 7-2. The familiar traditional pattern of a parquet floor.

Figure 7-3. Multiple colors of wood can create more interesting patterns.

Figure 7-4. Angles other than 90 degrees provide more possibilities.

Figure 7-5. Antique flooring reproduced from Wikimedia Commons.

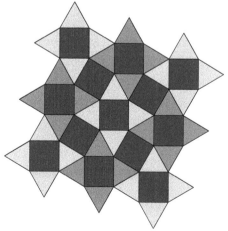

Figure 7-6. A parquetry pattern that can be made from plywood on a small scale.

Am I suggesting that you can lay your own parquet floors? No, I'm thinking on a smaller scale: using parquetry to decorate a trivet mat with a pattern of squares and triangles shown in Figure 7-6. I think you can tackle this, if you have practiced your plywood-cutting skills in Chapter 6. (And for those who don't know, a trivet mat is a mat that you place under a serving dish or pan, to stop it from burning the table.)

I should warn you that this is the most difficult project so far, because it requires you to cut 29 small pieces with precision. If you start on it and decide it's more than you want to deal with, there are many traditional parquetry patterns that are simpler (although maybe a bit less interesting). I have included a few at the end of this section, on page 89.

Fabrication Plan

The easiest way to make this project is to use some high-quality birch plywood that you can find in a crafts store. After some experimenting, I decided that ⅛" birch was easiest to work with, provided it is clamped firmly to a sacrificial piece to reduce the tendency for splintering.

Natural birch is quite pale. By applying a dark stain to some pieces and a brown stain to others, you end up with three colors.

After cutting and staining a bunch of squares and triangles, you will glue them onto another piece of plywood, which serves as a foundation. Trim the edges, and you're done.

YOU WILL NEED

- Two-by-four pine, any condition, length 18"
- Plywood, birch, ⅛" thick, size 12" x 12"
- Wood stain, brown, minimum quantity
- Epoxy glue and hardener, 2 oz
- Protractor, plastic
- Sandpaper, 220 grit

Also, as listed previously: Tenon saw, trigger clamps, ruler, speed square, rubber sanding block, work gloves, awl, dust mask (optional), safety glasses (optional), utility saw (optional), plywood work surface, sandpaper, polyurethane, disposable gloves, and three paint brushes (optional).

Check the Buying Guide on page 248 for information about buying these items.

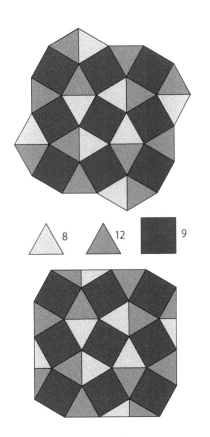

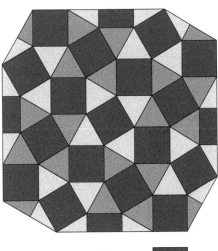

Figure 7-7. Design for a trivet mat.

Figure 7-8. For those who are willing to cut more plywood.

Pattern Options

The pattern at the top of Figure 7-7 is taken from Figure 7-6 and can be trimmed to form the shape below it. The straight outside edges will be much easier to cut than the zig-zag edges created by the geometrical shapes. If you're more ambitious, the version in Figure 7-8 is a possibility.

To create triangles and squares for these patterns, you can use vector-graphics software such as the application I described in Chapter 6.

If you prefer to use a pencil and a ruler to draw your squares and triangles, I have good news. It's a little-known fact that an equilateral (equal-sided) triangle with a base 30mm long is almost exactly 26mm high. (I'm describing it in millimeters for the same reason I have used millimeters to measure plans in previous projects—they are easier to mark out with a ruler.)

If you use pieces of this size, the pattern in Figure 7-7 will be about 4½" x 4½", which will be a relatively small mat, and the pattern in Figure 7-8 will be about 7" x 7". You can reduce the size of the first mat to make a coaster, by using smaller pieces, but they'll be difficult to cut accurately with a tenon saw. Miniature saws are available for people who build models, but they're outside the scope of this book.

Cutting Options

You have two cutting options for this project. First is the "on-the-line" option. This entails centering the saw blade on each line and cutting on it so that the thickness of the blade removes the same amount on each side. This is the same system that I suggested in Chapter 6, but in this project it is more of a challenge, as the pieces are smaller and the lengths of their sides must be equal.

Option 2 is the "outside-the-line" option, in which you cut outside of every line. This requires less skill, but will take more time because more cuts are required.

To cut triangles for the on-the-line option, you start with a plan shown in Figure 7-9. You can draw it on paper, then tape it to the plywood and prick through the intersections of the lines. Alternatively, it can be drawn directly onto the plywood. The upper part of the figure shows parallel guidelines 26mm apart, with marks every 15mm. The lower part of the figure shows how you join the marks to create triangles.

Figure 7-10 illustrates the process of cutting on each line. The red strips illustrate the width of the saw blade; all of the wood under each red strip will be removed. You begin by cutting horizontal strips, then cut along each remaining diagonal line in each strip. The resulting triangles are shown at the bottom.

If you prefer the outside-the-line option, Figure 7-11 shows that once again you begin with horizontal guidelines marked at 15mm intervals and spaced 26mm apart. But each pair of guidelines is separated vertically from the next by 10mm. The second part of this figure shows how you space the triangles, so that you will be able to cut outside the perimeter of each one.

Figure 7-12 illustrates that once again, you begin with horizontal cuts to create strips of triangles, although this time two cuts are required for each strip. Then remove each triangle from each strip, again cutting outside the lines.

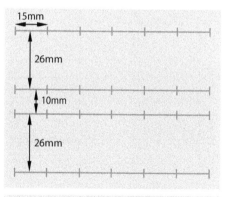

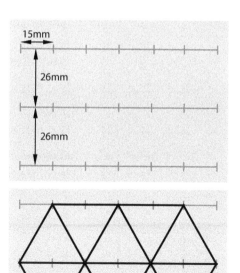

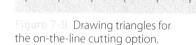

Figure 7-9. Drawing triangles for the on-the-line cutting option.

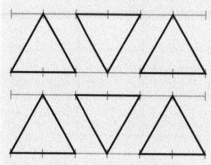

Figure 7-11. Drawing triangles for the outside-the-line cutting option.

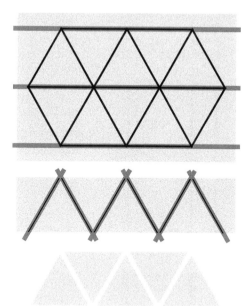

Figure 7-10. Cutting on the lines. The red areas indicate the width of the saw blade.

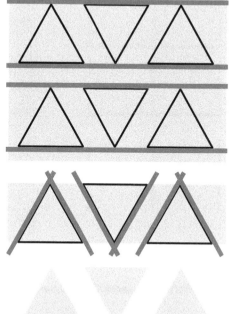

Figure 7-12. Cutting outside the lines.

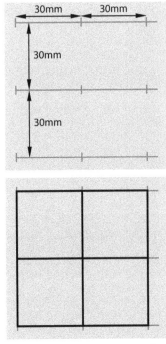

Figure 7-13. Drawing squares for the on-the-line cutting option.

Figure 7-14. Cutting on the lines.

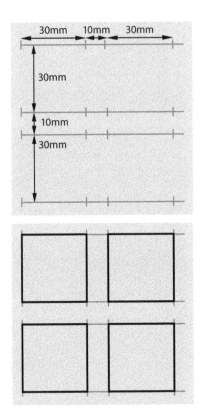

Figure 7-15. The plan for cutting outside the lines defining each square.

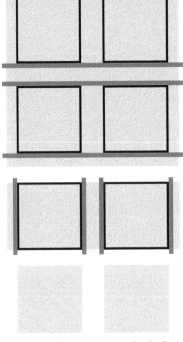

Figure 7-16. Cutting outside the lines.

Whichever option you choose, your system for drawing squares has to match. If you make triangles by cutting on the lines, but you make squares by cutting outside the lines, their sizes will be different, and the shapes won't fit together.

Figure 7-13 shows the plan for on-the-line square-cutting. Note that the guidelines are now 30mm apart.

Figure 7-14 shows that you begin by cutting the squares in strips, after which you cut each one out of each strip. Once again, the red lines indicate the width of the saw blade.

Figure 7-15 shows the plan if you want to cut squares with the outside-the-line option. This is similar to the plan for triangles.

Figure 7-16 once again shows that you cut the squares in strips, after which you cut each one out of each strip, except that this time you are cutting outside the lines.

Whichever option you use, print or draw the plan onto paper, so that you can transfer it to the plywood. If you draw the plan by hand, measure inward from the edges of the paper.

I have to warn you that because the pieces are so small, this project is quite difficult. You can make it easier if you are willing to spend a few dollars for a pull saw of the type shown in Figure 1-43 on page 17, because the thinness of the blade and the fineness of the teeth are ideal for this project. Still, with some care and patience, it should be possible with a tenon saw.

Personally, I used a tenon saw *outside* the lines.

You will need a guide piece on top of your plywood and a sacrificial piece under it for all the accurate cuts. The sacrificial piece makes long cuts difficult, as the angle of your saw will entail cutting a substantial groove in it. I suggest you work on a piece of plywood no more than 6" wide. If you begin with a piece that is 12" x 12", as in Figure 7-17, start by making a rough cut down the middle without worrying about splintering.

Now you can transfer strips of squares and triangles from your paper version onto the wood. When you start cutting, your job will be easier if your sacrificial piece and your guide piece are also 6" long. Figure 7-18 shows the setup for cutting the first edge of one strip. Remember, I chose to cut outside the guidelines.

In Figure 7-19, the plywood has been turned around for cutting outside the second guideline.

In Figure 7-20, a strip is ready for removing individual triangles. To continue cutting outside the guidelines, the strip must be rotated for each new cut.

After you cut the pieces, you will need to stain appropriate numbers of them with the two colors shown in Figures 7-7 and 7-8. Test the colors on a scrap piece of plywood before you use them on the finished project.

First smooth the surface with fine sandpaper (around 220 grit), being careful not to round the edges or the corners. Remove all the dust with a damp cloth, and allow the wood to dry thoroughly. Then apply the stain with a rag or wadded paper towel. Some people prefer to use a brush,

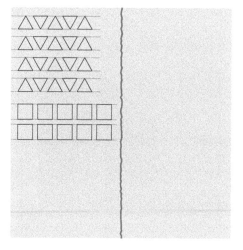

Figure 7-17. Your work piece for cutting triangles and squares should be no wider than 6". (The remainder of your 12" x 12" square of plywood can be used to make a base for your trivet mat.)

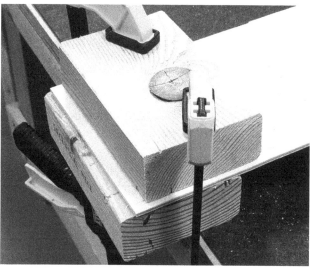

Figure 7-18. Preparing to cut a strip from the plywood.

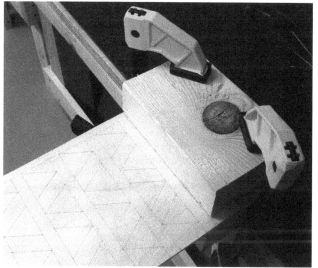

Figure 7-19. Ready to cut the second guideline.

Figure 7-20. Cutting individual pieces out of the strip.

but the opinions seem evenly divided about this. The important part is to wipe off any surplus stain within a few minutes. Then consider whether you want to apply a second coat, bearing in mind that this will darken the colors. See page 74 for additional notes about wood stain.

The stain should dry within an hour or so.

Finally cut a foundation for your pieces about 6" x 6", and you're ready for some gluing.

All About Epoxy

Carpenter's glue is probably not the best choice for this project, because it should really be clamped, and you have so many pieces, you'll have a hard time clamping them equally. I suggest you use epoxy glue, which does not require such rigorous clamping.

▪ Note that some types of epoxy are relatively toxic, and you don't want to get it on your skin. Wear disposable gloves, and use adequate ventilation.

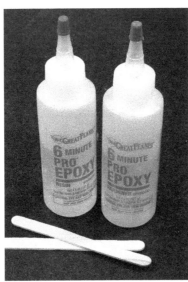

Figure 7-21. Plastic bottles of epoxy glue and hardener.

Epoxy is a type of resin which solidifies when it reacts with a separate liquid known as a hardener. When you buy epoxy, the resin and the hardener are packaged separately. They may be in individual bottles, such as those in Figure 7-21. Smaller amounts are often packaged conveniently in a dual reservoir from which the liquids are dispensed by pressing a single plunger.

You squeeze a little puddle of each liquid onto a piece of scrap wood or cardboard, and then mix them together with a **spatula**. This is not the kind of spatula you find in a kitchen; it's just a piece of thin wood like a popsicle stick. You should be able to buy a bag of spatulas from the same source as the glue.

Most forms of epoxy require equal quantities of resin and hardener, but some use a different ratio. Check the package before using it.

In Figure 7-22, the resin and hardener have been squeezed onto a piece of scrap wood. The next step is to stir the liquids vigorously with a spatula, to make sure they are thoroughly mixed. Then apply a dab to each piece of the trivet and set it on the plywood that you are using as a foundation. The assembled pieces are shown in Figure 7-23.

Epoxy is sold in versions that set quickly or less-quickly. Thirty-minute epoxy gives you time to shift everything around before it sets, but my impatient nature makes me prefer the six-minute stuff.

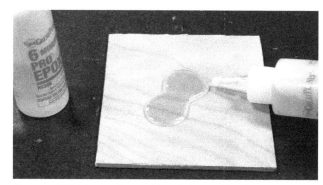

Figure 7-22. Mix equal quantities of glue and hardener.

Whichever type of epoxy you use, it takes several hours to reach its maximum strength. It will look as if it has set, but don't stress it until it has been undisturbed for a while.

Epoxy has several advantages:
- It works on almost any material.
- After it sets, it resists most chemicals.
- It's very strong.
- You don't have to clamp it hard, so long as it makes good contact with the materials that you are gluing.
- A thick layer works almost as well as a thin layer, unlike most glues.

Figure 7-23. The pieces are now glued to a base layer of plywood.

But of course, it has some disadvantages:
- It costs more than carpenter's glue.
- It's not as simple to use.
- It will be very difficult to remove from clothing or any other place where you don't want it to be.

Regarding the cost, this varies a lot from one source to another. See the Buying Guide on page 248 for suggestions.

Some points to remember:
- When the hardener makes the epoxy set, this is a chemical reaction that may take longer at low temperatures (below around 50 Fahrenheit). The warmer it is, the faster it sets.
- After the epoxy has set, there is no way to reverse the process.
- Never get the caps for the two bottles mixed up! If you put the cap for the hardener on the bottle of resin (or vice-versa), you may never get it off again.
- You can't clean up epoxy with a wet rag, as you can with carpenter's glue. You can scrape the drips with a razor blade, or sand some smears to get rid of them.

Figure 7-24. Ready to trim the edges of the design.

After the epoxy has hardened completely, you can apply a coat of polyurethane. Dab it on with a pale-colored rag or paper towel, and don't wipe it vigorously across the pieces. The oil-based polyurethane can pick up some stain pigment and will spread it to adjacent pieces if you rub it to and fro, ruining the multicolored effect. Dab it cautiously, and check frequently to see if you are picking up pigment. (This is why, if you use a rag, it should be pale in color.) Refold to expose a clean area of the rag or towel if necessary.

After the polyurethane has dried thoroughly, you may want to sand it with fine sandpaper (220 grit or higher) and apply a second coat.

When the glue has completely hardened and the polyurethane is completely dry, you can trim the design using a guide piece clamped above the mat, as in Figure 7-24. A sacrificial piece is underneath, not visible in this picture.

Maybe you would prefer to cut the base to match the zig-zag contours of the squares and triangles, but that would not be easy. Straight cuts are the best option, even though they divide some of the shapes in your parquetry.

Apply stain to the sawn edges, then smear some epoxy on all the edges, to seal them. The finished mat is shown in Figure 7-25.

Tesselations

Sometimes two different worlds come up with strangely similar ideas. Parquetry is really a kind of **tesselation**, which is the geometrical process of dividing a flat surface into shapes that interlock, leaving no spaces between them.

Figure 7-25. The finished mat.

The great master of tesselations was M. C. Escher, a Dutch artist whose work became known in the 1960s. I can't reproduce any of it here, for copyright reasons, but I encourage you to search for his name online. You'll find a lot of his art on Google Images—and you'll see that some people have reproduced it in the form of marquetry.

Other Projects

If you want to try something that involves fewer pieces and only 90-degree
and 45-degree angles, there's no shortage of traditional parquetry designs. I'm
reproducing four of them in Figure 7-26. I leave it to you to copy or redraw them and
figure out the best ways to cut the pieces.

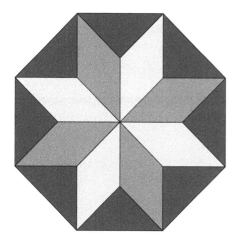

Figure 7-26. Some traditional parquetry designs.

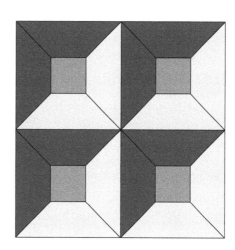

The simplest design of all consists of squares, with no other shapes at all. Use 64
squares and stain half of them dark, and you have a chess board. You already know
how to add a frame around it. This would be a simpler project than the trivet mat, but
less interesting and more time consuming, because of the large number of squares.

Chapter 8
Drilling It

NEW TOPICS IN THIS CHAPTER

- Choosing and using a drill
- Understanding drill bits
- Cutting clean, vertical holes

YOU WILL NEED

- Two-by-four pine, no knots, not warped, length 7"
- One-by-six pine, no knots, not warped, length 6"
- Finishing nails, 1¼"
- Electric drill and drill bits (see text for details)
- Countersink, ½", uniflute

Also, as listed previously: Tenon saw, trigger clamps, ruler, speed square, rubber sanding block, work gloves, awl, hammer, pliers, dust mask (optional), safety glasses (optional), plywood work surface, and sandpaper.

Check the Buying Guide on page 248 for information about buying these items.

I have not yet dealt with a very fundamental operation in fabrication: drilling holes in the materials that you are using. I'm going to suggest a very simple project that requires fairly accurate drilling—as soon as I tell you about the tools that you need.

Drills and Bits

A **drill** has a motor in it, which turns a **chuck**. The chuck holds a **bit**. Figure 8-1 identifies these parts.

The most common kinds of bits are **drill bits** and **screwdriver bits**. Samples are shown in Figure 8-2.

A drill bit is the thing that makes holes. Its tip is beveled with cutting edges. Spiral **flutes** extract wood particles from the hole that you are drilling. A smooth section above the flutes is called the **shank**, and fits into the jaws of the chuck.

Figure 8-1. The principal parts of a cordless drill.

A screwdriver bit is like the tip of a screwdriver. I'll be dealing with screws and screwdrivers in Chapter 10.

Confusingly, some catalogues and other sources may refer to drill bits as **twist drills**. Even more confusingly, in everyday life, someone may say "Pass me those drills," actually referring to a set of drill bits.

In this book, I'm using this terminology:
- A drill is the tool with a motor in it and a chuck on the end.
- A bit fits into the chuck. It may be either a drill bit or a screwdriver bit.
- A drill bit makes holes.
- A screwdriver bit fits into screws.

You can learn more about drills and bits by reading the Drills and Drill Bits Fact Sheet on page 101. But just in case you don't own a drill, I'll discuss your buying options here.

Choosing a Drill

Once upon a time, electric drills for personal use did not exist. For home repairs, you would use a hand-cranked device of the type shown in Figure 8-3.

You can still buy hand-powered drills online, but I'm assuming you prefer something easier. You need an electric drill, which is likely to be your biggest expense among all the tools required for the projects in this book. So how much money do you need to spend?

If you don't know how seriously you will be using tools in the future, I think you should spend as little money as possible. There's no need for a premium-quality product unless you will be using it for years to come.

This still leaves you with several choices, the first of which is whether your drill should be **cordless** or **corded**.

A corded drill plugs into a power outlet, and is cheaper than a comparable cordless drill—in fact, I just saw one advertised in the Harbor Freight catalog for the price of a McDonald's Happy Meal. (The Harbor Freight catalog is a well-known bargain-basement source for tools, if your primary consideration is finding a low price.)

Corded drills have the advantage that they don't require recharging, and they don't depend on expensive batteries that eventually require replacement. On the other hand, power cords tend to get in the way, even when you are working near an electrical outlet. They can knock things over, and may trip you. And if you're not within range of an outlet, you'll be additionally encumbered with an extension cord.

If you do decide to buy a corded drill, you don't need the macho, high-powered kind, with a handle that sticks out at the side (shown in Figure 8-4 on page 93). This will be unwieldy and more expensive than a more lightweight drill, and the extra power is unnecessary unless you are planning to do heavy work such as drilling holes in concrete.

Figure 8-2. A drill bit (left) and screwdriver bit (right). Many variations exist.

Figure 8-3. A manually powered drill.

Chucks and Batteries

Regardless of whether you decide to be corded or cordless, you should know about the two types of chuck.

To tighten the chuck, some drills require you to insert a **key**, consisting of a lever and a gear. This type of chuck can apply a very firm grip to the shank of a bit, but for the projects in this book, you should be happy with a **keyless chuck**, which you can loosen and tighten with your fingers (while the drill is not running, of course).

If you buy a cordless drill as opposed to a corded drill, the same Harbor Freight catalogue that I mentioned before offers a battery-powered model for the price of two Happy Meals. That sounds a good deal—but, you have to read the small print. This model uses a **NiCad** battery, abbreviated by the catalogue as NiCd. (The term "NiCad" is trademarked, but is used generically, like "Scotch tape.")

A NiCad is much bigger and heavier than a **lithium-ion** battery that delivers the same power, and its lifetime will be limited if it isn't treated carefully. It should not be left sitting in a charger for excessive periods, and should not be recharged until it is fully discharged. If you don't use a NiCad for two or three months, some sources recommend "exercising" the battery by discharging it and recharging it. Bearing this in mind, you should only use a NiCad if you don't mind the extra weight and can maintain it conscientiously.

A lithium-ion battery is much more tolerant of erratic charging and discharging habits, but a drill with this kind of battery may cost as much as three Happy Meals, or even four.

Sourcing Your Drill

After you decide what you want, check for special offers at big-box stores (especially if there's a national holiday), or on eBay, where you can buy second-hand drills cheaply. You may want to avoid second-hand battery-powered drills, because you won't know if a NiCad battery has been properly treated, and even a lithium-ion battery has a limited number of charge-discharge cycles.

Often, a drill is bundled with an electric screwdriver, and they may be packaged in a nice carrying case. This seems a tempting proposition, as a screwdriver will be mandatory in future projects, and a manual screwdriver becomes tiresome quite quickly.

Beware, however, of sets that only contain one battery. If you have to swap a battery between a drill and a screwdriver, you might as well just own a drill, and swap drill bits and screwdriver bits.

Choosing Drill Bits

You can buy a full set of drill bits ranging in size from ¹⁄₁₆" to ½" in diameter, in increments of ¹⁄₆₄". For our purposes, this is overkill. You don't need all those sizes for the projects in this book, and affordable handheld drills tend to be limited to a maximum shank size of ⅜". (Remember, ⅜" is smaller than ½".)

I think 11 sizes of drill bits should be sufficient (listed below), and if you buy them as a set, they should cost no more the drill. At the time of writing, DeWalt offers what I consider to be an ideal limited set. The sizes are:

- ¹⁄₁₆" x 2
- ⁵⁄₆₄" x 2
- ³⁄₃₂"
- ⁷⁄₆₄"
- ⅛" x 2
- ⁵⁄₃₂"
- ³⁄₁₆"
- ⁷⁄₃₂"
- ¼"
- ⁵⁄₁₆"
- ⅜"

Figure 8-4. You don't need a heavy-duty drill for the projects in this book.

The set includes two bits of sizes ¹⁄₁₆", ⁵⁄₆₄", and ⅛", because these are the ones that tend to wear out, get broken, or get lost before the others.

Whatever brand you choose, bear in mind that the projects in this book only require you to drill holes in wood or soft plastic, and therefore you don't need to buy exotic bits. They don't need to be tungsten-coated (although, if they are, that's fine). You don't care whether they have a black finish or a bright metal finish, and you should not be concerned by the angle of the cutting edges at the tip.

However, you should be careful not to buy drill bits calibrated in millimeters—or masonry bits, which are not suitable for wood.

See the Drills and Drill Bits Fact Sheet on page 101 for additional information about drills and bits.

Using a Drill

Looking at the drill from the front, and assuming that your drill has a keyless chuck, you turn the chuck counter-clockwise to open its internal jaws to accept the drill bit that you want to use. Insert the bit and tighten the chuck, making sure that the bit is properly centered. Note that some larger bits often have flat spots on the shank, which must be aligned with the jaws of the chuck to make a good fit.

Before pulling the trigger on a drill, you need to know which way the bit is going to turn. Almost all modern drills are reversible, but the location of the lever or button to control this will vary depending on the manufacturer. Check the instruction leaflet that came with the drill, and also remember that there may be a center, locking position of the lever in which the drill will not respond.

The speed of a drill usually varies depending how far you pull the trigger. Many drills also have a two-position switch to choose a speed range. For drilling wood with a bit that is ¼" or smaller, the higher speed range is appropriate. If you are drilling a larger hole, the bit will tend to get hot if you run it fast, especially if you are using it in hardwood. You don't want it to scorch the wood, and when you remove the bit from the drill, you don't want it to burn your fingers.

To test your drilling technique, insert a relatively small bit, such as ⅛", and use it on a piece of scrap two-by-four. But please read the following safety precaution first.

Drill Safety

A drill seems harmless, except that—well, in some ways, it isn't.

If you have long hair, loose shirt cuffs, a necklace, a necktie, or anything else that dangles into your work area, it can get caught and trapped in the rotating drill bit, and you'll discover how surprisingly powerful a little handheld tool can be. The results will be painful. Tie back your hair, button your cuffs, and don't wear anything that could get caught. See Figure 8-5.

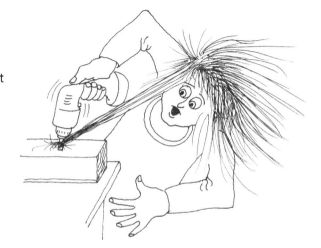

Figure 8-5. Long hair must be tied back in a workshop. Always!

Another rule is never to drill into a piece of wood that you are holding with your hand, especially if your hand is wrapped around the wood. A drill can make a hole in your hand with no trouble at all, and a visit to the nearest emergency room will be in your future. See Figure 8-6.

You should clamp any object that you are drilling, and two clamps are better than one. This is especially important

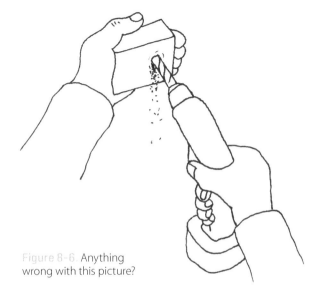

Figure 8-6. Anything wrong with this picture?

when using a drill bit that is ¼" or larger. The bit can grab the wood, and if you are holding the wood in your hand, the bit can pull it out of your fingers. A long piece of plywood can start spinning like the blade in an electric fan, and can hurt you.

Be alert for the tendency of a drill bit to skitter away from the place where you want it to go. If you are holding the drill with one hand, keep your free hand at a safe distance. Remember that every drill bit terminates in beveled sections that are probably sharper than the knives in your kitchen.

Gloves may seem to provide an additional layer of safety, but the fabric of the glove can get caught in the rotating bit, possibly injuring your finger. For this reason, OSHA now recommends that employees should not wear gloves while using a drill.

Drugs and alcohol are always a bad idea in a workshop, but especially when using power tools. Safety glasses are generally a good idea, as a drill can disperse fragments unpredictably.

One last cautionary note. It's natural to pick up a drill by its handle, and when you do, it's easy for your finger to press the trigger. A cordless drill is "always on," even if you are merely carrying it from one place to another. If the bit starts rotating unexpectedly, it can surprise you. You can avoid this risk by getting into the habit of pulling out the battery—or if the drill has a forward-reverse switch with a center, locked position, you can leave it locked.

Making a Clean Hole

Sometimes you want to drill a hole all the way through a piece of wood, in which case you must clamp it on top of a sacrificial piece, as in Figure 8-7. This will not only protect your work area but will help to prevent splintering where the drill bit emerges from the wood that you are drilling.

If you are sure that a hole won't go all the way through a piece of wood, you can clamp it as in Figure 8-8.

When the bit penetrates deeper into wood, it can get clogged, preventing it from functioning properly. Pull the bit out, and you'll find compacted wood particles near the tip. You must clear the flutes, perhaps by scraping them with a thin piece of wood, bearing in mind that the metal may be hot from friction.

Design for a Bit Holder

Now, finally, it's time to get into your first drilling project. A set of drill bits is usually sold in a clamshell case that doesn't allow for quick and easy access. A bit holder solves this problem. It can be simply a block of wood in which you drill a hole for each bit, after which you put each bit in its hole.

Figure 8-7. When drilling all the way through a piece of wood, clamp a sacrificial piece below it!

Figure 8-8. If you are absolutely positive that you will not drill all the way through a piece of wood, you can clamp it like this.

Of course, it isn't quite as simple as that, and the process of completing this very basic project will be instructive.

Figure 8-9 shows a simplified rendering of my design. It requires just 7" of a two-by-four and slightly less than 5" of a one-by-six.

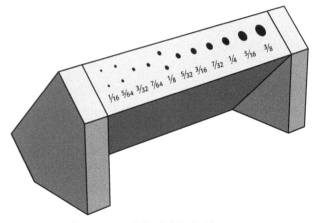

Figure 8-9. The concept of the drill-bit holder.

Figure 8-10 shows my suggestion for arranging the set of 14 drill bits that I mentioned on page 93 by inserting them into the edge of your two-by-four. If you have more bits, or fewer bits, you'll need to modify the plan. Either way, you should print it onto paper and transfer it to your piece of two-by-four by pricking through the paper with an awl.

Two-by-four pine (viewed from the edge)

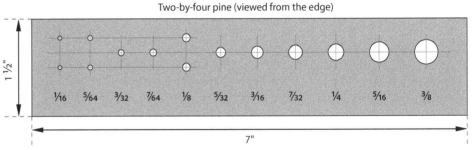

1 ½"

| ¹⁄₁₆ | ⁵⁄₆₄ | ³⁄₃₂ | ⁷⁄₆₄ | ⅛ | ⁵⁄₃₂ | ³⁄₁₆ | ⁷⁄₃₂ | ¼ | ⁵⁄₁₆ | ⅜ |

7"

Figure 8-10. A possible layout for 14 drill bits (including three duplicates).The precise locations of the holes is up to you to decide.

Drilling Procedures

You now come to the first problem associated with drilling wood, which is that in softwood such as pine, the pale rings in the grain are much softer than the dark rings in the grain. Therefore, a drill bit will always try to wander into a pale area.

One way to deal with this is to create a deeper, wider guide hole with your awl by using more muscle power. You can hammer the awl, if necessary. The hole that you create should be so well-defined, it will tempt the drill bit to go there instead of going where it would otherwise prefer to go.

The next problem you will encounter is that when you are drilling soft wood, you will not only get splinters around an exit hole (if you go all the way through the wood). A large drill bit will also create splinters where it *enters* the wood.

Figure 8-11 shows what I mean. This splintery mess was created by using a ⅜" bit without making any preparations.

Here's the remedy that I suggest. First you will use your awl, creating a hole like the one near the bottom of Figure 8-12.

The next step is not widely used, but it works for me. You can use a **countersink** to widen the awl hole.

A countersink is a type of bit that is really intended to bevel a hole for the head of a wood screw. But you can use it here to prepare the way for a drill bit. Because it has a sharper cutting angle than a bit, it makes a relatively clean, cone-shaped hole, as shown in Figure 8-13. (The countersink is in the chuck of the drill, in this photograph. It is a **uniflute** type of countersink, meaning that it has only one flute and cutting edge.)

Next you use a bit that is smaller than ⅜". I tried one that is ³⁄₁₆", as shown in Figure 8-14.

Finally, in Figure 8-15, a ⅜" bit has enlarged the hole, with relatively little collateral damage. The countersink has left a small bevel around the edge of the hole, but personally I like the way this looks.

Note that if you use hardwood, splintering is much less of a problem.

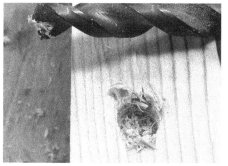

Figure 8-11. What happens when you start right in with a ⅜" drill bit on soft pine.

Figure 8-12. Your awl should make a guide hole that looks like this.

Figure 8-13. A countersink can enlarge the hole made by your awl, and the edges will be relatively clean.

Figure 8-14. It's always easier to drill a smaller hole, and then enlarge it, than to go straight for a large drill bit.

Figure 8-15. Finally, you have a reasonably clean hole in your soft pine.

Completing the Drilling

Having figured out how to make a clean hole, the next challenge is to make it at 90 degrees to the surface of the wood. Here again, wood grain can divert the bit. You should pause and check the angle of the drill from the front and from the side, to make sure that it hasn't deviated.

Some drills have a **bubble level** built in—meaning, a little capsule containing liquid with a bubble in it. When the bubble is centered, the drill is vertical and the hole that you are drilling is at 90 degrees relative to the horizontal surface.

Always keep a firm grip on the drill, preferably with two hands, and always clamp the wood, preferably with two clamps, because the large bit can get a good enough grip to make the wood rotate unexpectedly if it is only anchored at one point.

If you are making a deep hole (which you will, in this project), don't forget to pull the bit out occasionally to clear it of impacted chips.

Adjusting the Fit

I will now assume that you have completed your mission to drill a hole for every bit. The next question is, do you want to store your bits with the sharp end up or the sharp end down? I think inserting them with the sharp end down makes them easier to grab. However, you'll find that the same bit which created a hole doesn't slide very easily into the hole. It's a tight fit. How do you loosen it?

The easy answer is to widen each hole by using the next size larger bit. This will be okay so long as the holder doesn't tip over, allowing the bits to fall out.

The ⅜" bit is a problem, though, because you don't have a larger bit than this. So, what you can do is run the drill fast and push the bit in and out of the hole repeatedly, while leaning the drill a little to each side. This is not considered good practice, especially with smaller bits that can bend or break, but it's something that most people do anyway. The flutes of a bit have hardly any cutting power, so don't expect a quick result.

Fabricating the End Pieces

The plan for the end pieces is in Figure 8-16. Start by cutting a section of one-by-six that is 4⅝" long, with the grain oriented as shown in the figure.

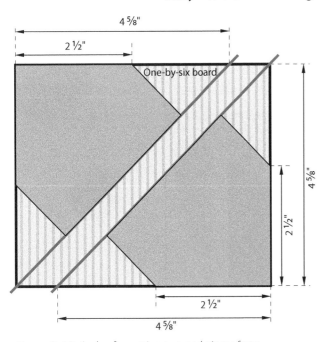

Figure 8-16. A plan for cutting two end pieces from a small piece of one-by-six pine.

Cut along the two red lines, first. Because these lines are angled at 45 degrees, you can draw them with your speed square, as shown in Figure 8-17.

Using a piece of sacrificial wood may be difficult, unless you precut it to match the triangular shape that you are cutting. But I'm inclined to think that a sacrificial piece may not be necessary for this job, so long as you cut at a shallow angle.

After you cut along the two red lines, you need to make four more cuts to get the two five-sided brown shapes shown in Figure 8-16. I think all of these cuts can be done freehand, as in Figure 8-18, although your miter box may be just big enough for the five-sided pieces, if you prefer.

To attach the end pieces to the two-by-four, you can use some of those 1¼" finishing nails left over from previous projects. But here's an idea: to reduce the risk of splitting the one-by-six, and to be sure that the nails will go where you want them to go, you can drill a little hole for each nail with a ¹⁄₁₆" bit. These are properly known as **pilot holes**, because they guide the nails into the wood.

This project seemed like a no-brainer, but I'll bet you ran into a few unexpected issues. In particular, I'm betting that some of the holes you drilled turned out to be at odd angles. How could that happen? While you were making the holes, surely the drill seemed properly aligned.

Don't feel bad if this happened to you. Keeping a handheld drill precisely vertical is really quite difficult—which is why you may eventually decide that you need a drill press. I will have more to say about that on page 243.

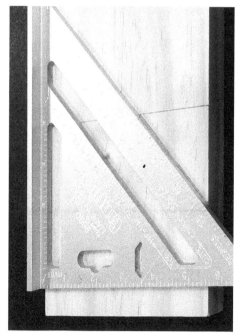

Figure 8-17. Use your speed square to create 45-degree-angle lines.

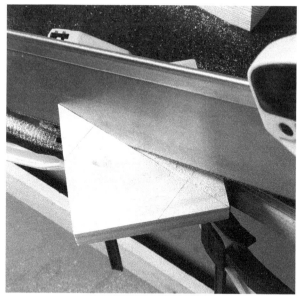

Figure 8-18. The end pieces can be cut freehand. Splintering should be minimal if you hold the saw almost parallel with the wood when it is nearing its exit point.

Meanwhile, my version of the bit holder is shown in Figure 8-19, with bit sizes that I printed onto a strip of paper. Do you think this item needs a coat of polyurethane? Maybe not. For an object in a workshop, bare wood may be good enough.

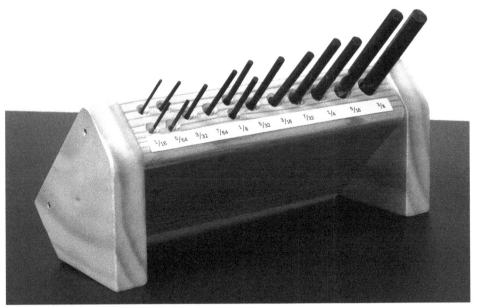

Figure 8-19. Finished drill-bit holder.

Other Ideas

What else can you build, using your newly acquired ability to drill holes in wood? Well, a cribbage board certainly has a lot of holes in it. So does a Chinese checkers board, or the kind of solitaire board that has a cross-shaped pattern of holes with pegs in them. You could even make a traveling chess set, if you can find chess pieces that will fit into the holes.

Drills and Drill Bits Fact Sheet

A typical specification for a drill should provide you with the following information:

Torque

Torque is a measure of turning force. Obviously you would like as much torque as possible, but the specifications for drills don't always give you a number that you can compare between different models. A corded drill (which plugs into the wall) is potentially capable of delivering more torque than a cordless drill (which is battery powered).

Weight of the Drill in lbs

Lighter is generally better, for precise work with small objects.

Maximum Speed

Usually around 1,200 rpm. You don't need more than this. A second, low-speed range is desirable, and its maximum speed may be listed. Assuming that the speeds are set by internal gearing, the low range should have more torque than the high range, and should be greater than the torque of a single-range drill. Merely running a single-range drill more slowly by pressing its trigger more lightly will not give you more power.

Maximum Shank Size

Often ⅜" but occasionally ½". The larger capacity is good to have, and will probably mean that the drill can deliver more torque, even if it is not stated. A ½" drill is likely to be more expensive than a ⅜" drill.

Battery Capacity

This will be in **amp-hours**, abbreviated Ah. It tells you how many hours you can run the drill, theoretically, while it is drawing 1 ampere of current. You can use this number to compare the battery capacities of similar drills.

Voltage

A higher **voltage** generally can deliver more power. Most drills are either 12V (relatively low, these days) or 18V to 20V (average).

Hammer Drill Capability

When the drill can't deliver enough power to force the bit more deeply into some hardwood, it switches automatically to a mode where it turns the bit in little pulses. This is a desirable feature. It is also useful for drilling holes in masonry.

Clutch

You can adjust this by turning a ring, to stop the drill when it applies a certain amount of torque. The idea is that if you use your drill as a screwdriver, you can set the clutch so that it starts to slip before the drill applies enough force to snap the screw. Not all drills have this feature. (Mine does, but I never use it.)

LED Light

Some drills include an **LED** that illuminates the spot immediately in front of the drill, so that you can see what you are doing in low-light situations. The LED switches itself on when you start drilling, and may stay on for 10 or 20 seconds after you let go of the trigger. When I first saw this, I thought it was a gimmick; but after using it, I like it a lot.

Which Manufacturer?

Everyone has a favorite, but this tends to coincide with the product that someone just bought for a lot of money. The more you spend, the more motivated you are to feel that it was a wise investment. Personally I like DeWalt products, and I have friends who like Milwaukee, Makita, Bosch, and Hitachi. If you aren't going to be using a drill frequently, you may be satisfied with less-expensive brands.

Drill Bits Basics

High-speed steel bits are for general use in most materials. **Cobalt-steel** bits are for very hard materials such as stainless steel, titanium alloys, and nickel alloys. **Carbide-tipped** bits have better wear resistance in abrasive materials such as fiberglass and aluminum. **Solid carbide** bits are harder, stronger, and offer greater wear resistance than high-speed steel, cobalt steel, and carbide-tipped bits, but are brittle and must be handled carefully. Special-purpose bits are available for drilling hard materials such as tile, glass, and masonry. For the projects in this book, cheap bits will do the job, because no challenging materials are involved.

Length

Options are **short** (for tight spaces), **jobber's length** (average), and **extended reach**. The extra length of an extended-reach bit may be essential if you need to drill a hole through a two-by-four or two-by-six inside a wall—for example, to install a wired internet connection.

Auger

An **auger** functions like a regular bit except that it has a screw tip and an open-spiral design, able to remove more debris. Also the auger uses horizontal cutting blades instead of angled blades at its tip. It is more expensive than a typical drill bit.

Stepped Bit

This consists of a cone-shaped stack of cutting rings that gradually increase in size as you move from the tip up. Two examples are shown in Figure 8-20. Because these bits were lightly greased, they retained some debris that is visible.

The idea is that a single bit can drill multiple hole sizes, just by pushing it in further. The big disadvantage is that it only works when drilling materials that are thinner than the step size of each ring.

Figure 8-20. Two stepped bits.

Another Type of Stepped Bit

The term *stepped bit* is also used to describe a bit that has two sections, one to create a pilot hole for the thread of a wood screw and a wider hole for the smooth shank of the screw. (You'll learn about wood screws in Chapter 10.) The two sections may connect with a small step, or the bit may taper from one diameter to the other, as shown in Figure 8-21. This is properly known as a **taper-point** drill bit. Note the angled black countersink section at the top, to bevel the edges of the hole. This section can be loosened with a hex wrench for relocation up or down the bit.

Figure 8-21. A taper-point drill bit with countersink at the top.

Larger Holes

Special bits to drill larger holes are discussed in the Holes and Curves Fact Sheet on page 164.

Chapter 9
A Swanee Whistle

NEW TOPICS IN THIS CHAPTER

- A different kind of jig
- Precision drilling
- How a whistle works

YOU WILL NEED

- Two-by-four pine, no knots, not warped, length 10"
- Round dowel, hardwood, ¾", length 18" to allow for errors
- Round dowel, hardwood, ⅜", length 12"

Also, as listed previously: Tenon saw, miter box, trigger clamps, ruler, speed square, rubber sanding block, work gloves, awl, utility knife, electric drill and drill bits, countersink, dust mask (optional), safety glasses (optional), plywood work surface, sandpaper, and epoxy glue and hardener.

Check the Buying Guide on page 248 for information about buying these items.

A penny whistle is usually made from metal tubing, but with enough care, you can make something similar out of wood. The instrument you build in this project will actually be a Swanee whistle, which was invented in England in the 1800s and is sometimes called a slide whistle. Your version will be very high-pitched, because of the limitations of your drill set. You'll have to adjust it to make it work—but in the end I think you will get it to whistle.

This project requires you to drill a long, straight hole with a ⅜" bit. You already did that in the previous project, but this one will be a little more difficult, as the hole will be going down the center of a round hardwood dowel.

You'll need to be careful while the drilling is in progress. If the drill slips, or if it splits the dowel open, you don't want your hands to be nearby. Make sure the dowel is clamped securely, and keep both hands on the drill while you are operating it.

Drilling a Round Dowel

Some round dowels are made from softwood. The lines of the grain are clearly visible, looking like a two-by-four. This is not what you want. The grain should not be so easily visible, and any label on the dowel that you buy should identify it as being a hardwood such as oak, maple, or poplar (not pine).

Start with a piece of ¾" round dowel that is 6" long. This is much longer than your ⅜" drill bit, but you're going to drill a hole from each end, meeting in the middle. Can it be done? I think so, if you are careful and methodical.

The first step is to decide how to clamp the dowel. You could lay it flat and clamp it to the surface of your work bench, but then you would have to drill into it horizontally. I have difficulty making a horizontal hole that doesn't drift from side to side, so I'll suggest that you clamp it vertically.

To do this, you need a jig. I'm thinking of using two pieces of two-by-four, each 3" long. You will cut a V-shaped channel in each of them, and they'll grip the dowel between them, preventing it from rotating when you're drilling a hole in it. The finished version is shown in Figure 9-6 on page 106.

To minimize the risk of the jig splitting while it is being used, you should cut each V-shaped channel across the grain.

Figure 9-1 shows how the wood should be marked, to guide you in making the saw cuts.

The two lines on the surface of the wood can be made by using your speed square, at 90 degrees to the front edge of the wood. Their distance from each end of the wood is unimportant, but the lines must be ¾" apart, parallel with each other.

After you draw them, use your speed square to extend them over the edge at 45 degrees to the surface of the wood, as shown in the figure.

How will you make these angled cuts? Your miter box may have a slot, usually located at one end, that allows the saw to lean over at 45 degrees relative to horizontal. I don't recommend this, as I have never found it easy to use.

Figure 9-2 shows how you can set things up to make the cut without a box.

Don't imagine that you can start right in with the saw at an angle; it will slip. The way to do it is to make an initial vertical cut, about ¹⁄₁₆" deep, as shown in Figure 9-3.

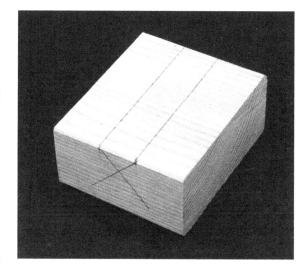

Figure 9-1. A 3" piece of two-by-four marked for sawing. The parallel lines are ¾" apart. Note the direction of the grain.

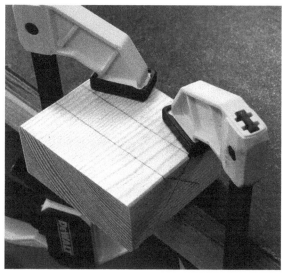

Figure 9-2. The wood ready for cutting.

Figure 9-3. Make an initial vertical cut to stabilize the saw.

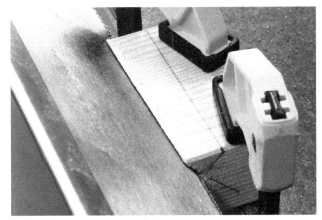

Figure 9-4. Making the angled cut.

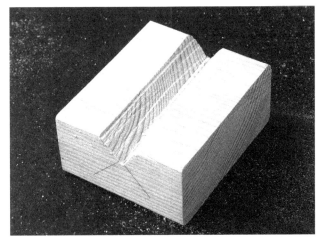

Figure 9-5. One piece of the jig completed.

Figure 9-6. The two halves of the jig, completed.

When you have a groove across the wood, you can lean the saw over as in Figure 9-4, and start the angled cut.

▪ Notice that the teeth of the saw are close to the bars of the clamps. Be very careful that your saw doesn't rub across them. The teeth will be blunted, and the saw will never be the same again.

Figure 9-5 shows one of the wooden blocks with a V-shaped groove completed. The sides of the V are messy, but they are parallel with the sides of the block, which is the main thing. Small irregularities won't prevent the groove from clamping the round dowel.

Cut the second block, exactly the same as the first. They can stand opposite each other as in Figure 9-6.

Now you can place the 6" piece of ¾" dowel in the jig as shown in Figure 9-7. I stood a couple more blocks on either side, for extra stability. One clamp is applied horizontally, to hold everything together. A second clamp is applied vertically, to stop the whole assembly from moving when you start to drill into it.

As described in Chapter 8, I think it's a good idea to proceed in small steps when you're drilling a ⅜" hole. This is less important when you're dealing with hardwood than when you're drilling into pine, but small steps enable you to correct errors as you go along.

The first step is to use an awl to mark the center of your dowel. Finding the center can be tricky, but you can use your ruler to check that the point of the awl is at an equal distance from each side. Then turn your ruler and check again from other directions.

After you make a hole with the awl, put a countersink in the drill and create a cavity as shown in Figure 9-8. You should do this a little at a time, pushing the drill to one side or the other if the cavity isn't quite centered.

Now use a ¼" bit to start a vertical hole. If your drill doesn't have a bubble level to tell you that it's vertical, you should pause frequently and check the bit from the front, the side, and then the front and the side again, as it goes deeper. Don't press too hard, and remember to pull the bit out occasionally to get rid of the accumulated wood chips. The more slowly you proceed, the better your chances are of getting it right.

Remember that the spiral flutes on a drill bit are necessary to excavate wood particles from the hole that you are drilling. Therefore, you don't normally drill beyond the extent of the flutes. If the smooth, round shank of the bit enters the hole, it will start to get hot quite quickly.

In Figure 9-9, after I made a hole as deep as possible with the ¼" bit, I stood a screwdriver in the hole to check it was vertical.

Now for the interesting part. Switch to your ⅜" bit and run the drill slowly. Touch the bit to the ¼" hole very gently. Making the ⅜" hole is a very different experience from making the ¼" hole. The bit will really grab the wood if you let it. Be cautious and don't hurry.

Your goal is to drill downward for just over 3", which a typical ⅜" bit should be able to manage without any trouble. Clean out all the debris, and the result should look something like Figure 9-10. Not perfectly centered, but close.

Now you need to loosen the clamps and take the jig apart, to clean out all the fragments of wood. You don't want any debris in the way when you tighten the jig again.

Turn the dowel upside-down, clamp it, and repeat the drilling sequence from the other end. I doubt that the two holes will meet precisely, but they don't have to, so long as they overlap to some extent.

You may find that when you break through from one hole to the other, the bit will grab the dowel and may start turning it, no matter how tightly you clamped it. This is less likely if you are keeping the flutes of the bit clean, but if it does happen, do not hold the dowel to stop it from turning! There's a risk that the dowel may split open, at which point you will find yourself holding the drill bit inside it. That would be a very unpleasant experience.

If the drill jams, put the drill into reverse, pull the bit up, then return to forward gear and increase the speed. You should be able to make the transition from one hole to the other inside the wood.

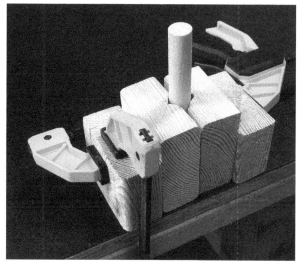

Figure 9-7. Ready for drilling.

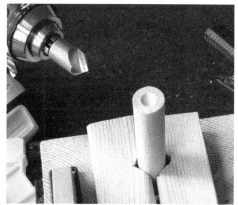

Figure 9-8. A cavity created with a countersink will make drilling a hole much easier.

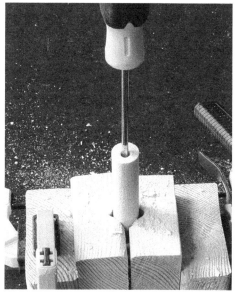

Figure 9-9. A screwdriver in the ¼" hole verifies that it was drilled vertically.

Figure 9-10. Part one of the task completed.

Figure 9-11. Making a vertical cut that will become the top end of the fipple hole in your whistle.

Figure 9-12. Beveling the fipple hole with a utility knife. Be sure to cut away from you. Keep your other hand away from the cutting site.

No doubt you'll want to unclamp the dowel and look through it to make sure the holes are aligned—in which case, you'll probably be disappointed. The ⅜" bit will have pushed the last bit of sawdust into the lower hole, instead of extracting it. You may need to use a screwdriver or pencil to clear the dowel.

Blow through it, but be careful not to inhale when your mouth is close to the dowel. You don't want to breathe sawdust. Turn your head to take a breath of clean air, then blow through the dowel. If it still isn't clean inside, you'll need to apply the drill again.

Making the Mouthpiece

A Swanee whistle is closed at the bottom, and all the air will emerge through a hole that you cut into the side of the dowel. The part of the whistle which you blow into is called the *fipple*, and the hole where the air emerges is the *fipple hole*. (That's a real word—I didn't make it up. In fact you can find entire web sites online, dedicated to discussion of fipple geometry in penny whistles. It's not a trivial matter.)

Getting the fipple to work is tricky, because tiny variations make a big difference, but if you're patient, I think you'll get a sound out of it. I've completed this project three times, and I always managed to make it emit a high-pitched whistle in the end.

The first step is to make a saw cut 1" from the end of the dowel, with the jig turned on its side to hold the dowel horizontally, as shown in Figure 9-11.

Be very gentle with this cut. You don't want it to go too deep. As soon as the saw reaches the hole inside the dowel, stop immediately.

Now you need to make a diagonal cut to meet the vertical cut, and I think the easiest way is with a utility knife, as shown in Figure 9-12. Keep the dowel clamped on the bench and push the knife away from you, keeping your other hand out of the way.

The hole has to be as clean as possible. Ragged edges will diffuse the air stream and prevent your whistle from resonating. Use the knife very delicately to trim the edges of the hole, as in Figure 9-13.

The next step is to restrict and direct the air flow through the mouthpiece and under the hole in the side of the whistle. To do this, you need to plug the mouthpiece with a 1" length of ⅜" dowel that has a flat side, as shown in Figure 9-14.

You can sand a piece of ⅜" dowel to make a flat spot on it. This is most easily done as in Figure 9-15, before you cut the 1" section.

Insert the ⅜" dowel partially into the upper end of your whistle, as shown in Figure 9-16. Don't push it in all the way. Blow into the end with the plug in it, so that your breath goes through the thin gap between the plug and the hole that you drilled. Block the the opposite end of the dowel with your finger. If all you get is a vague, breathy sound, move the plug a fraction further in or further out. Still no note? Sand the plug some more and try again. The dimensions are critical, and the force with which

Figure 9-13. Trimming the edges of the hole. The knife is now facing you, so use it very gently, and do not apply much force.

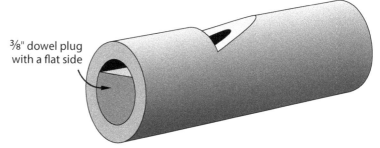

⅜" dowel plug with a flat side

Figure 9-14. The basic principle of fipple design is to restrict the air flow and direct it under the hole in the side of the whistle.

Figure 9-15. A flat spot on the ⅜" dowel, created by sanding it.

Figure 9-16. Adjust the depth of the plug and test the whistle repeatedly.

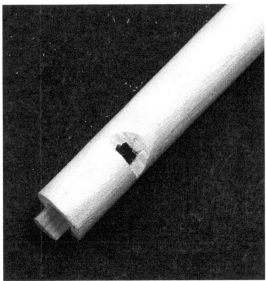

Figure 9-17. A beveled mouthpiece on the whistle.

you blow will also have some effect. You shouldn't need to blow hard. In fact, you are more likely to get a sound if you blow very gently.

Be careful not to inhale while your mouth is over the dowel. The plug should fit tightly, but just in case it is loose, you want to avoid the risk of drawing it into your mouth.

When you finally have a whistling sound, you can cut the plug so that it is flush with the end of the dowel, and glue it in place. Then sand it to make a better-shaped mouthpiece. See Figure 9-17.

Finally, cut a 3" piece of your ⅜" dowel and slide that into the other end of the whistle, as in Figure 9-18. By moving it in and out, you'll vary the whistling pitch.

A larger whistle would create a deeper, louder, mellower tone. How could that be made? Well, instead of wood, how about using a plastic tube, such as a piece of PVC water pipe? I'll begin to explore the whole topic of plastics in Chapter 15.

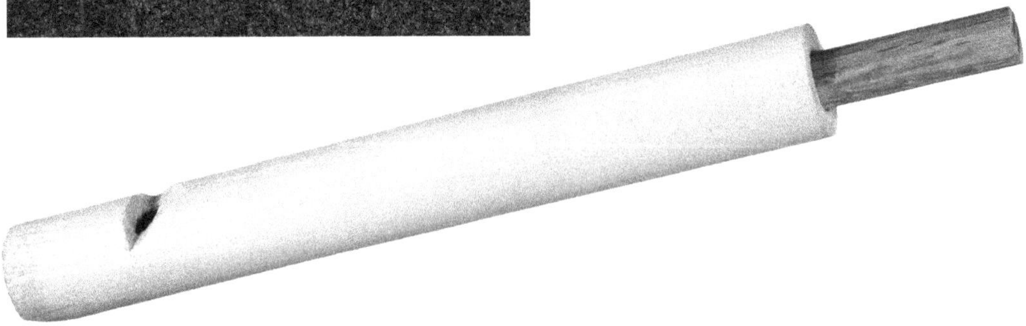

Figure 9-18. The finished whistle.

The concept of a screw thread is more than 2,000 years old. The great Greek inventor Archimedes developed a massive wooden screw as a method for raising water from a river. Much later, in the 1800s, screws and bolts became an essential part of the industrial revolution. The modern world, as we know it, would not work without them.

This project uses screws where glue wouldn't be as strong, and it introduces you to design issues associated with one of the world's most basic shapes: a rectangular box.

You'll find facts about screws in the Screws and Screwdrivers fact sheet on page 123. I'll also mention the option to buy an electric screwdriver, but for this project, you only need a manual screwdriver with a medium-size (number 2) Phillips head.

A Screw and a Square Dowel

Before I get into the project itself, I want you to do a little wood-splitting experiment, like the one when I discussed nails in Chapter 4. You need a 3" piece of ¾" x ¾" square dowel (maybe left over from previous projects), a 3" piece of ½" x ½" square dowel, a screwdriver, and three wood screws. They can be 1" long, #10 size, as shown in Figure 10-1, although any #10 screws will do.

The #10 classification describes the thickness of the **shank** of the screw. In a wood screw, the shank is the smooth part directly under the head (like the smooth section of a drill bit). Unlike nails, which tend to be fatter when they are longer, a #10 screw is the same thickness no matter how long it is. You'll find a table of screw sizes in Figure 10-34, in the fact sheet at the end of this project.

Figure 10-1. A screwdriver tip and screw for a wood-splitting experiment.

Begin with the piece of ¾" square dowel. How can you drive the screw into it? Well, you can just press hard, turn the screwdriver, and hope that the screw will cooperate—but this is not easy, especially if your dowel is hardwood.

NEW TOPICS IN THIS CHAPTER

- Types of screws
- Types of screwdrivers
- How to design in 3D
- Types of corner joints

YOU WILL NEED

- Two-by-four pine, any condition, length 18"
- Square dowel, hardwood, ½" x ½", length 18"
- Square dowel, ¾" x ¾", length 3"
- Plywood, ¼" thick, size 12" x 18"
- Wood screws, #6 x ⅝", flat head, Phillips, quantity 20
- Wood screws, #10 x 1", flat head, Phillips, quantity 3 (optional)
- Screwdriver, Phillips size 2, or electric screwdriver and bits (optional)

Also, as listed previously: Tenon saw, miter box, trigger clamps, ruler, speed square, rubber sanding block, work gloves, awl, masking tape, utility knife, electric drill and drill bits, countersink, dust mask (optional), safety glasses (optional), utility saw (optional), plywood work surface, sandpaper, and epoxy glue and hardener.

Check the Buying Guide on page 248 for information about buying these items.

You can whack the screw with a hammer to embed the tip of it in the wood, but a better idea is to use an awl. Set the screw aside, and push the sharp tip of the awl into the wood, about ¾" from the end. Lean on it hard, and wiggle it around to make a cavity.

Figure 10-2. What happens when you force a #10 screw into a small piece of ¾" x ¾" dowel.

Now you should find that starting the screw will be no problem, but you'll need to use more strength as it goes in deeper, and I'm betting that the wood will split, as in Figure 10-2. Screws, like nails, do tend to split wood, especially if the screw size is relatively thick.

A #10 screw is really too big for this job. But there's a way to make it penetrate a hardwood ¾" dowel if you really want to. You drill a **pilot hole**.

The Importance of a Pilot Hole

When a drill makes the hole, it doesn't compress wood significantly. It extracts some of the wood as tiny chips. The hole then provides space for the body of the screw, while the threads of the screw, which do the real work, cut into the material that remains around the hole.

Try using a ⅛" drill to make a pilot hole for another #10 screw. If you're wondering why I chose ⅛", a larger pilot hole would not allow the screw to get a firm grip, while a smaller pilot hole might still allow the screw to split the wood. You'll find a table of recommended hole sizes in Figure 10-34.

Now when you insert the screw you should find that it slides in more easily and does not split the wood, but is still secure. With a ⅛" pilot hole, you can even embed a #10 screw in a skinny little ½" square dowel, as in Figure 10-3.

In Chapter 8, when you were building a drill rack, I suggested the option of using pilot holes with finishing nails. If you are using screws in small, precise projects, pilot holes are not just an option anymore. They are essential.

Figure 10-3. If you drill a ⅛" pilot hole, a #10 screw can even fit into a ½" x ½" square dowel without splitting it.

Corner Blocks

The current project is for a basic box, because the shape of a box is fundamental in fabrication work. Each of the drawers that slide out of your kitchen cabinets is constructed as a box shape. A cupboard is basically a box. Your chess set may be packaged in a wooden box. Even a bookcase is box-shaped.

Your first, basic box will be ultra-simple. The sides, bottom, and lid will be made from plywood that is ¼" thick. Small blocks of square dowel, ½" x ½", will connect the sides at the corners.

Figure 10-4 shows a rendering of the design for this box with its front side and its lid missing. Screws will be driven through the plywood and into the blocks inside the corners.

There are many other ways to make corners, and I'll list some of them at the end of this project. But blocks are the easiest option.

Cutting Plywood

I designed this box to be small, to minimize your sawing, but I will assume that you may be starting with a relatively large piece of plywood.

When I visited my small-town local lumber yard to see what they had available, they offered me some shopworn, slightly scratched ¼" plywood with a soft pine veneer that looked as if it was guaranteed to splinter. Could I really build a small box with that? I decided I should find out, just in case you ever have a similar experience. They sold me a quarter-sheet measuring 48" x 24".

Even if you buy better-quality plywood, it often has ragged or worn edges. Therefore, it is standard procedure to make your own edges by cutting inside the edges created by the lumber company.

Figure 10-5 shows the first step. The ragged red lines indicate rough cuts, without worrying about splintering the underside of the wood, because you can't easily use a piece of sacrificial wood under a cut that's 16½" long.

Incidentally, here's a quick tip:

If your saw tends to jam or squeal while you are making a long cut, take hold of the free end of the wood and twist it up and to the left, away from the cut, as shown in Figure 10-6. This will relieve the pressure on the saw blade. Just be careful to keep your fingers away from those saw teeth.

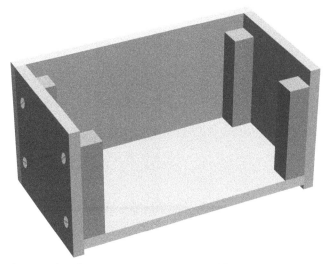

Figure 10-4. Box construction using corner blocks.

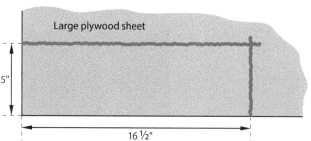

Figure 10-5. The first two cuts are rough cuts.

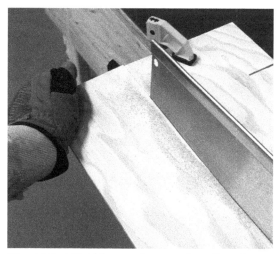

Figure 10-6. Bend the wood up away from the cut to relieve the pressure on the saw.

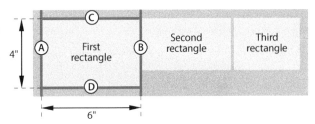

Figure 10-7.
The first of three
rectangles from
your work piece.

Now that you have a manageable
work piece, you can extract three small
rectangles from it. Figure 10-7 shows
the cuts that will give you the first
rectangle. I'll get to the second and the
third in a moment. My idea is that none
of the cuts should be longer than 6",
so that you can make them easily with
your tenon saw.

First make a clean cut along line A,
about ½" inside the edge, and roughly
parallel with it. The precise position is
not crucial. Use sacrificial wood under
the plywood as in Figure 10-8.

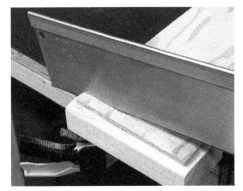

Figure 10-8.
Cutting along
line A.

After you make a nice straight cut, you
can use this as your **reference edge**.
This means you will make the next lines
and cuts with reference to it. All the
measurements for this project will be
made relative to that edge, because
the other edges were either rough-
cut or factory-cut, and are not entirely
trustworthy.

Now that you have your reference edge,
make two marks on it that are 4" apart,
as shown in Figure 10-9.

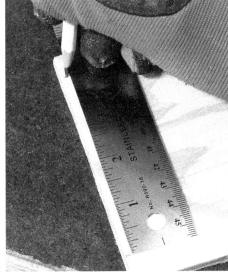

Figure 10-9.
Two marks, 4"
apart, on line A.

Place your speed square against the
reference edge, and draw the two
6" lines that were labeled C and D in
Figure 10-10. Use your pencil to make
a mark at each 6" location.

Draw line B between the two 6" marks
that you just made, and then extend
line B out to the side edges of your strip
of plywood, as shown in Figure 10-11.
(Refer back to Figure 10-7 if you've
forgotten which line is line B.) You
just drew a rectangle based on your
reference edge. So long as your speed
square is accurate, your rectangle
should be accurate.

Figure 10-10.
Measure 6" along
the lines that were
labeled C and D in
Figure 10-7.

Cut along line B, as shown in Figure 10-12. When the cut is done, set aside the remainder of your work piece, rotate the rectangle that you are working on, and lines C and D are now short enough to be cut edge-to-edge with your tenon saw. Remember to cut outside the pencil lines at each step.

Retrieve the remaining portion of your work piece, which will enable you to make your second rectangle. In Figure 10-13, you'll see that line B is your new reference edge. Make two marks on it, 3" apart. Then you can use your speed square to draw lines for cuts E and F, each 5½" long. Then draw line G and cut along it. Finally cut E and F. This is the same system that you used for the first rectangle.

In Figure 10-14, you see the third step in your procedure. When you complete it, you should have the three rectangles shown in Figure 10-15.

You see, now, why I didn't suggest making a bigger box. There's quite a lot of cutting involved. But you're not finished yet: you need to make a second, identical set of three rectangles (because a box has six sides, assuming it includes a lid). Retrieve your original sheet of plywood and repeat the steps beginning with the rough cuts shown in Figure 10-5.

After you have your six pieces of plywood (two of each shape), you need four pieces of ½" x ½" square dowel, each 3¼" long. You can use the miter box to cut these sections.

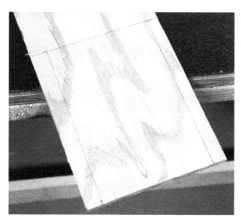

Figure 10-11. The perimeter of the first rectangle has been outlined.

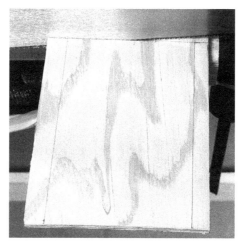

Figure 10-12. Cutting along line B.

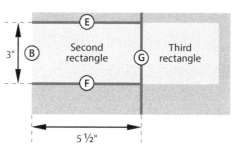

Figure 10-13. Laying out the second rectangle.

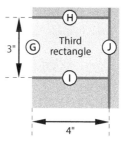

Figure 10-14. Obtaining the third rectangle.

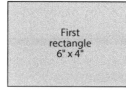

Figure 10-15. The three rectangles.

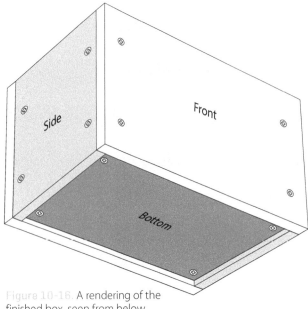

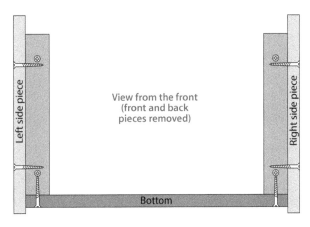

Figure 10-16. A rendering of the finished box, seen from below.

Drilling the Plywood

Figure 10-16 shows your ultimate objective: the box, seen from below. I used contrasting colors for the front, side, and bottom pieces, to match the colors in Figure 10-17, which shows two X-ray views revealing the locations of the screws. You can see that it's important to avoid having screws running into each other from different directions.

Don't worry if you find the plans confusing; everything will make sense when you assemble the pieces.

The locations for screw holes in the five sides of the box are shown in Figure 10-18, Figure 10-19, and Figure 10-20. (The sixth side of the box is the lid, which is not screwed into place, because it is removable.)

You can print or draw the plans on paper, and then transfer them to the plywood pieces by pricking through with an awl. Alternatively, you can make measurements directly on the wood, but you'll still need to prick the drilling locations with the awl.

Use a $5/32$" bit to make the holes in the plywood, which have to be big enough for the shank of each screw. These holes go all the way through, so clamp a sacrificial piece of wood under the plywood to prevent splintering where the drill bit breaks through.

When you're starting each hole, run the drill very slowly and press the bit very lightly. A small drill bit should not normally chew up the plywood, but if your plywood has a pine veneer as soft as mine, it can happen.

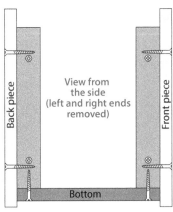

Figure 10-17. An X-ray view revealing the locations of screws.

If you want to do the best possible job, you can prepare for each hole with a countersink, using the technique that I described on page 97 in Chapter 8. But bear in mind, you also need to use the countersink on each hole after you drill it, to bevel it so that it will accept the angle of the screw head. Apply the countersink until the bevel is just a little wider than the diameter of a screw head.

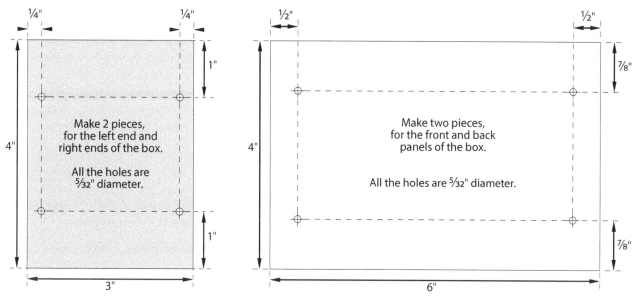

Figure 10-18. Screw hole locations for the ends of the box.

Figure 10-19. Screw hole locations for the front and back of the box.

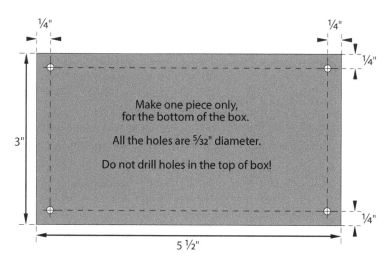

Figure 10-20. Screw hole locations for the bottom of the box.

Make four dowels as blocks
for the corners of the box.

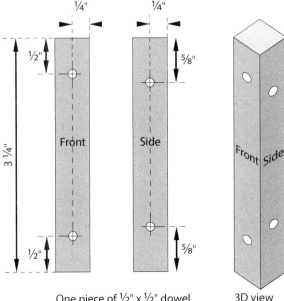

Figure 10-21.
Pilot hole locations
in a dowel.

One piece of ½" x ½" dowel
viewed from two adjacent sides.

Each hole is ³⁄₃₂" diameter
drilled ³⁄₈" deep.

The two other sides of the dowel,
not shown here, do not have
any holes drilled in them.

3D view

Figure 10-22. Drilling pilot holes, with
masking tape around the ⁵⁄₆₄" drill bit
showing the depth limit.

Drilling the Dowel

In the pieces of dowel, you will use a ³⁄₃₂" bit to create pilot holes. You only drill into two sides of each dowel, as shown in Figure 10-21. I included an extra 3D view in case the plan isn't clear.

The holes do not go all the way through the dowels, and must be ³⁄₈" deep. Put a little piece of masking tape around your drill bit at the ³⁄₈" mark, as shown in Figure 10-22, so you will know when you have reached the depth limit.

This is not the only way to measure and align pilot holes, but I'd like you to do it this way in this project. The reasons will soon be clear.

Don't bother to make pilot holes in the ends of the square dowels.

If you are wondering why the screw holes are larger than the pilot holes, Figure 10-23 provides an explanation. The screw threads are not supposed to grip the plywood (labeled A in the diagram). They grip the dowel, and the head of the screw pulls the plywood down against the dowel, while the smooth shank at the top of the screw rotates freely. This is why wood screws have a smooth section at the top. (Or at least, they used to. I've noticed that some so-called wood screws are now being made without the smooth section.)

The box under assembly is shown in Figure 10-24. When you attach each dowel, make sure that it's the right way around, so that its remaining holes are visible and ready for the next piece of plywood. Because each dowel is symmetrically drilled, you can turn it upside-down to expose the remaining holes if necessary. If the sides of the box are not quite aligned with each other, you can sand down the high edge later.

After you make the front, back, and sides, you may have to sand the edges of the base of the box to make it fit. Once you have it in place, drive the screws in at each corner. I don't think you'll need pilot holes, as the screws are going into the end of the grain.

The lid may be tight until you sand it. Once you have it fitting neatly, I suggest a piece of round dowel as a handle, although a piece of square dowel will do just as well. Draw an X between the corners on the underside of the lid, to find its center. Then use one more screw to hold the handle in place—or glue it, as you wish. I happened to have a piece of ¾" round dowel left over from the Swanee whistle project, and I rounded the top of it just for fun. See Figure 10-25.

Imperfections

I'm guessing that your box will not be perfect. I think this is the most important lesson to learn from this project: no matter how careful you are, screws and drill bits tend to wander away from where they should be, tiny errors tend to accumulate, and the next thing you know, one side is higher than the other, or a guide hole doesn't align with a screw hole.

In future projects I'll be mentioning ways to minimize these errors, or to make adjustments as you go along. But you may be wondering if most of the measurements are even necessary. Can't you just put the screws approximately where they should go?

You can—but what will you do if something goes wrong, such as two of the screws bumping into each other inside one of the dowels? I think drawing a plan saves time in the long term, and if all the measurements are accurately made, the result looks better, too.

Vector graphics software is ideal for drawing plans. This is one reason I urged you to install OpenOffice Draw, back in Chapter 6. Very often, the software shows you errors that would have ruined your project if you started to build it without planning it properly.

When the screw is turned, its thread pulls it downward. The head of the screw pulls piece A toward piece B.

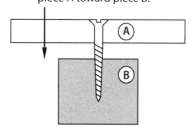
Figure 10-23. How a screw works.

Figure 10-24. Assembly in progress.

Figure 10-25. The box, completed.

Improvements in Corner Design

This box was an exercise, not a thing of beauty. Leaving aside the problem of building it precisely, how could the design be made better? In Chapter 16, I'll show how to make a better box out of plastic. But using wood, I see three areas for improvement.

1. The blocks in the corners are unattractive and occupy space.

2. Cheap plywood doesn't look nice.

3. The screw heads are visible, and they don't look nice, either.

Hiding the Screw Heads

Figure 10-26. Decorative chrome-plated round-headed screws.

One way to hide the screw heads is by covering the exterior of the box with **veneer**, so long as the heads are flush with the plywood, or slightly recessed. Veneer is a very thin layer of wood with an attractive grain pattern. You can attach it with contact adhesive. The same concept is used in kitchen counter tops, where particle board is covered with a laminate such as Formica.

But veneer is expensive, and is not easy to use.

You can, of course, recess the screw heads and then conceal them with caulk, plastic wood, or some other filler. Smear some more onto the edges of the plywood, and then paint the box. This would take time and trouble, and I suspect that joints between the sections might still be visible under the paint.

The most elegant solution is to frame each side of the box. The frame would consist of square dowels, each measuring maybe ⅜" x ⅜", with a groove cut in their inside edges. Very thin plywood would slot it into the grooves. I like the way this looks, but it isn't quick or easy.

An easy alternative would be to use decorative screws. Chrome-plated round-headed screws are shown in Figure 10-26. You wouldn't countersink the wood, if you were using these.

Other types of screws have a miniature threaded socket in the head. Dome-shaped or button-shaped accessories screw into the socket, transforming a defect (the screw head) into a feature (the dome concealing it). Screws of this type are sometimes used to hold a bathroom mirror onto the wall.

However, the best way to get rid of visible screws is by not using any screws at all. What are our options, there?

Materials

To hide the edges of the plywood at the corners, you could miter the corners. To avoid using screws, you could then glue the plywood to the interior blocks. But cutting and joining thin plywood precisely at 45 degrees is not easy, and in any case, the top edge of the plywood would still be visible.

The edges of natural wood are nicer to look at. While you have seen that thin wood breaks easily when flexed parallel with the grain, the sides of the box shouldn't be under much stress. Many options are available for corner joints using natural wood (see below).

Alternatively, you could get away from wood altogether, and use ABS plastic, making the corners by bending them. I'll be showing you how to do this in Chapter 15.

Losing the Blocks

If you switch to natural hardwood, and if you make the sides of the box thicker, there are dozens of ways to create corners. Here are some common ones.

Square-Ended Butt Joint

The most basic configuration simply butts one piece against another and glues them together. This isn't very strong, but is quick. See Figure 10-27. (I suggested this method for making the corners of the frame for Dad's Puzzler, in Chapter 1.)

Dowel Joint

Small round dowels, which I usually think of as **pegs**, can strengthen a butt joint. In Figure 10-28, pegs have been inserted in one panel, and will be glued into matching holes in the other panel. You'll find this method in the very next project, to build a bookcase.

Mitered Butt Joint

The joint that I have been referring to as "mitered" is really a **mitered butt joint**, meaning that the flat surfaces butt together. See Figure 10-29.

Mitered Reinforced Butt Joint

A groove can be cut in each mitered surface, and a thin rectangle of wood can be glued into it—assuming you have the right kind of tool to cut the grooves. See Figure 10-30. This type of reinforcement can be added to other types of joints.

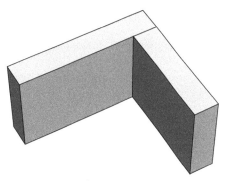

Figure 10-27. The most basic corner joint.

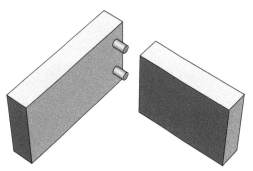

Figure 10-28. Dowels inserted to strengthen a square-ended butt joint.

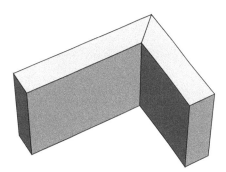

Figure 10-29. Mitered butt joint.

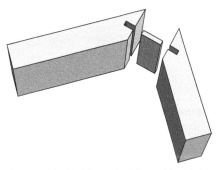

Figure 10-30. Mitered reinforced butt joint.

Rabbet Joint

Because glue is more powerful when it is applied to two surfaces at 90 degrees, a rabbet joint is stronger than a plain butt joint. See Figure 10-31.

Corner Lipping

A section can be added to the corner to conceal the edges of the wooden panels. The section can be rounded, as shown in Figure 10-32, or can have a square cross-section, or ornamental shape.

Many more joints exist, because people have been thinking of ways to deal with the basic problem of connecting pieces of wood at 90 degrees for hundreds of years. Search Google Images for board joints, and see how many you find. I'm not listing them here because many, if not most, require power tools or special hand tools to cut the necessary shapes in the wood.

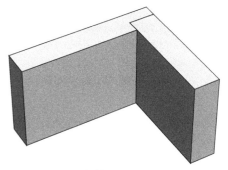

Figure 10-31. Rabbet joint.

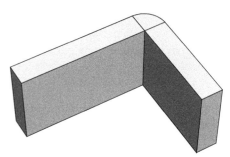

Figure 10-32. Lipping.

Screws and Screwdrivers Fact Sheet

Options for Screws

In a good hardware store, you'll find a bewildering variety of screws organized primarily by length, thickness, and type of head.

If the body of the screw has a smooth section near the head, it is a wood screw; if not, it is usually described as a sheet-metal screw. The smooth section is the **shank**, and is designed to make good contact with a hole in an upper, thinner piece of wood (or metal accessory), without gripping it. Sheet-metal screws are intended for use with thin metal, so they do not have a smooth shank. Their threads may be sharper. They are also suitable for many plastics.

Head Profile

Viewed from the side, the profile of a screw head usually falls into one of four main categories, although some other variants exist.

Flat head. Intended for use with wood or plastic in which a countersink is used to accommodate the beveled underside of the head, so that the head itself can be flush with the surface of the material.

Pan head. Does not require any countersinking. The head protrudes above the surface around it.

Round head. May be found securing wooden panels on boats, where it is likely to be made of brass. In general applications, it has been largely replaced by pan-head screws.

Hex head. The head is six-sided, for use with a **socket wrench** or **nut driver**.

Figure 10-00. Some sample screws. See text for details.

Some sample screws are shown in Figure 10-33. From left to right: square-drive sheet-metal screw; round-headed wood screw; hex-headed sheet-metal screw with additional screwdriver slot; generic pan-headed sheet-metal screw with Phillips head; generic flat-headed wood screw with Phillips head; hex-headed sheet-metal screw with self-tapping tip (makes its own hole); and a large carriage screw.

Length

In a flat-headed screw, length is measured from the upper surface of the head to the tip of the thread. In other types of screw, length is measured from the underside of the head to the tip of the thread.

In the United States, for general applications, length increases in ⅛" increments for short screws, and by ¼" when the screws are longer.

Thickness

The body of the screw is specified by a series of **gauge numbers**. A larger number indicates a fatter screw. Those that share the same number will have the same thickness, regardless of differences in length. Screws larger than #12 often have their thickness expressed in inches; thus #14 is equivalent to a ¼" screw, while #20 is usually sold as ⁵⁄₁₆", and #24 is ⅜".

Thread Spacing

In the United States, thread spacing is measured in **threads per inch**, sometimes abbreviated as **tpi** (not to be confused with teeth per inch, in a saw). The threads are usually more widely spaced on screws of larger diameter.

The table in Figure 10-34 shows gauge number, threads per inch, shank diameter, and recommended drill-bit size for a pilot holes in hardwood and softwood.

Screw Gauge	Threads per Inch	Shank Diameter (inches)	Pilot Hole (softwood, inches)	Pilot Hole (hardwood, inches)
#2	26	3/32	1/16	1/16
#3	24	7/64	1/16	5/64
#4	22	7/64	1/16	5/64
#6	18	9/64	3/32	7/64
#8	15	5/32	7/64	1/8
#10	13	3/16	1/8	9/64
#12	11	7/32	9/64	5/32
#14	10	1/4	5/32	11/64

Figure 10-34. Basic data for commonly used screw sizes. Information supplied by Jet machine tools.

Head Slot

A single, straight slot for a flat-bladed screwdriver is the oldest, traditional design. A cross-shaped head is properly known as **cruciform**, although few people use that term. It was introduced to satisfy demands of mass production, as it automatically centers the screwdriver and prevents it from skipping or sliding out.

In the United States, all cruciform screws tend to be described as having a **Phillips** head, although in fact there are variants. A true Phillips head has a small square cavity in its center, and the type of screwdriver that fits it should have a blunt tip, as shown in Figure 10-35. A cruciform screwdriver that has a pointed tip is really designed for **Frearson** screws, which do not have a small square cavity in the center. Using a Frearson screwdriver in a Phillips screw will tend to destroy the Phillips screw head, and vice-versa.

Figure 10-35. Three Phillips screwdriver bits, sizes 1, 2, and 3 from left to right.

Hybrid screw heads will accept a Phillips screwdriver, but one of the slots is extended to the edges of the screw so that it can also be turned with a flat-bladed screwdriver.

PoziDriv is another type of cruciform screw head, known as **PLC** outside of the United States, where Phillips screws are rare. The Pozidriv head, as its name suggests, provides a very secure fit, so long as it is used with a Pozidriv screwdriver.

Torx screws have a star-shaped head, and are gaining in popularity in the United States, as they provide a very secure grip. See Figure 10-36. Torx is a trademark, but the term is now used generically.

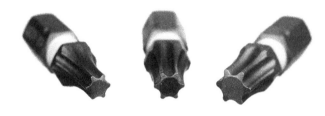

Figure 10-36. Three Torx screwdriver bits.

Square-drive screw heads have a square cavity. Compatible screwdriver bits are shown in Figure 10-37.

Some types of **tamper-proof** screw heads have a single slot that is beveled so that a straight-blade screwdriver cannot remove the screw, and can only tighten it. Other tamper-proof heads have a small peg in the center to prevent the most commonly used types of screwdrivers from engaging.

Figure 10-37. Three square-drive screwdriver bits.

Specific Applications

In addition to wood screws and sheet-metal screws, types have been developed for specific applications.

Drywall screws are designed to secure sheets of drywall when a building is being constructed. The underside of the head is curved to satisfy the characteristics of drywall. The screw is almost always black in color. The thread may be coarse, for rapid insertion into wooden framing, or fine, for use with a wall framed with metal studs.

Deck screws are thin wood screws, usually at least 2" long, for quick and easy insertion into softwood beams, without any need for a pilot hole. They are treated to resist corrosion, and are often gold in color. They are mostly sold in one thickness, roughly equivalent to #8.

Carriage screws, also known as **lag screws** and sometimes as **coach screws**, are large wood screws, usually with hex heads (sometimes, square heads). The fat body of a carriage screw is intended to carry substantial loads, but requires more torque than most battery-powered tools can provide. You are likely to use a socket wrench to tighten a carriage screw. Confusingly, carriage screws may be referred to as **carriage bolts**, **lag bolts**, and **coach bolts**, even though they are not bolts.

Machine Screws

A **bolt** is properly known as a **machine screw**, although the term is not used consistently. Nuts and bolts are dealt with in Chapter 14.

Manual Screwdrivers

Basic muscle-powered screwdrivers are often sold in sets, such as in Figure 10-38, which contains ⅛", ³⁄₁₆", and ¼" blade widths. (Two of each of the ³⁄₁₆" and ¼" blades are provided.) The very short screwdriver in this set is included for accessing screws in difficult locations. The double-ended, crank-shaped screwdriver has the same purpose, and is also very portable.

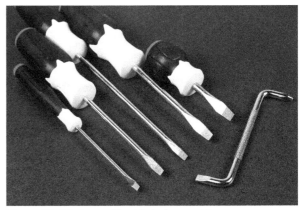

Figure 10-38. A set of flat-bladed screwdrivers.

A set of Phillips screwdrivers is shown in Figure 10-39, containing size 0, size 1, and size 2. Three screwdrivers of size 2 are provided, as this is the most commonly used.

A size 0 Phillips screwdriver may be too large to fit the miniature screws often used in electronics products. Sets of miniature screwdrivers can be bought for this purpose.

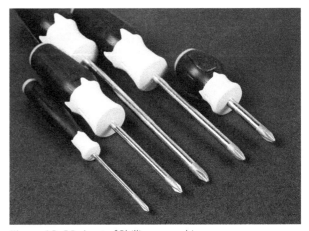

Figure 10-39. A set of Phillips screwdrivers.

Some manual screwdrivers have a shaft with a hollow tip that takes interchangeable bits of the type normally used in electric screwdrivers (described below).

Electric Screwdrivers

For the small projects in this book, a manual screwdriver is all you need, and can actually be an advantage when you want to avoid the risk of over-tightening small screws in soft material such as plastic. Bearing this in mind, I will not go into a lot of detail about specifications and buying options for electric screwdrivers.

I mentioned previously that you can use your drill as an electric screwdriver; all you need is a set of screwdriver bits, and possibly a bit extension to reach places where the body of the drill won't fit. You will spend some time tightening and loosening the chuck on the drill to switch between bits, but you may find this acceptable.

An electric screwdriver does not have a chuck. Instead it has a quick release mechanism, which is specifically designed for screwdriver bits and is quicker to use. A cordless electric screwdriver is shown in Figure 10-40.

If you do decide to buy a cordless electric screwdriver, you can choose one that is compatible with your electric drill, so that they can share the same batteries.

Figure 10-40 A cordless electric screwdriver with lithium-ion battery.

Chapter 11

A Bare-Minimum Bookcase

The Great Eastern Temple in Japan's Nara prefecture was built in the 8th century, rebuilt twice afterward, and remained the world's largest wooden building until 1998. You can see a photograph of it in Figure 11-1. Amazingly, all the wooden sections were held together with wooden pegs.

Figure 11-1. The Great Eastern Temple.

You, too can use pegs to join pieces of wood together—for example, when you build a small bookcase.

Problems with Store-Bought Shelves

I've never felt entirely satisfied with store-bought bookcases, or kits to make bookcases. There are three reasons:

 Although the shelves may be adjustable, they tend to average at least 10" apart. This is too much for many uses. When I make my own shelves, I can measure the items that I want to put on the shelves and design them to fit.

 In store-bought kits to make bookcases, the shelves and end pieces are usually made of particle board, which isn't very rigid, and they aren't thick enough, because thicker shelves would cost more money. As a result, the shelves tend to sag.

I don't like the look of shelves that rest on brackets inserted into tracks. I prefer a real bookcase, but bookcases are almost all designed to stand on the floor. A lot of wall space in my home is not used for anything above the six-foot line. It's a perfect place for shelving—but only if I build it myself.

This project will address these problems. The design will be small, but you'll be able to expand it significantly, if you wish.

Design Issues

By my definition, a shelf may or may not have an end piece, but a bookcase has two of them supporting the ends of the shelves. This raises the inevitable question: how should the shelves and the end pieces be joined together?

It's the same basic problem that I described when building a box, in which the sides had to be attached to each other at 90 degrees. In that project, I was stuck with the limitations of plywood. In this project, ¾" boards allow me to use a plain-and-simple butt joint that is reinforced with wooden **pegs**—also known as **plugs**, and sometimes as **dowels**, even though they aren't very similar to the dowels we have been using. To avoid confusion, I prefer to call them pegs.

You can buy pegs specifically designed for joining sections of wood. Your local hardware store may not have them, but you can find them in quantity online. They have grooves in them, or a rough, machined surface that is ideal for glue. Search for "grooved dowels," "fluted dowels," or "dowel pins," and you should find what you need. For this project, they should be ¼" diameter and 1¼" long, which is a popular size.

Here's the basic plan. First, you drill holes on the inside of each vertical support. The holes don't go all the way through. Then you glue the pegs into the holes, leaving about half of each peg sticking out. You drill matching holes in the ends of the shelves, and glue the shelves onto the pegs. (Some people prefer to do all the gluing at once, instead of in two stages. I find that a bit more challenging.)

Measuring and Planning

Before you start cutting wood, some decisions are necessary. How many shelves do you want? Perhaps three will be enough, just as proof-of-concept. Two of these shelves can be for books, while the top shelf can be used for ornaments. See Figure 11-2.

How long do you want your shelves to be? Because this is just a test project, I'll say 18", although the design will be horizontally scalable, so you can have shelves that are longer—even twice as long—if you prefer.

How tall should the end pieces be? Let's suppose you are just going to use this miniature bookcase for old-style paperback books that measure 7" tall and 4¼" wide. Now you can draw a plan to figure out the dimensions, using graph paper, or vector graphics software, or just a freehand sketch.

YOU WILL NEED

- Two-by-four pine, any condition, length 18"
- One-by-six pine, a few knots, not warped, length 96"
- One-by-two oak, maple, or poplar, length 40"
- Wood screws, #8 x 1½", flat head, Phillips, quantity 6
- Wooden plugs, ¼" x 1¼", quantity 12

Also, as listed previously: Tenon saw, miter box, trigger clamps, ruler, speed square, rubber sanding block, work gloves, awl, masking tape, electric drill and drill bits, countersink, screwdriver or electric screwdriver, dust mask (optional), safety glasses (optional), utility saw (optional), plywood work surface, sandpaper, polyurethane, disposable gloves, paint brush (optional), and nylon rope or thick string.

Check the Buying Guide on page 248 for information about buying these items.

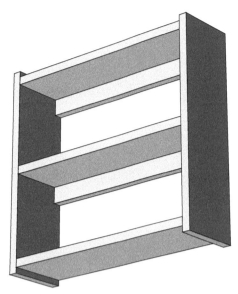

Figure 11-2. A rendering of the bare-minimum bookcase consisting of three shelves and two end pieces.

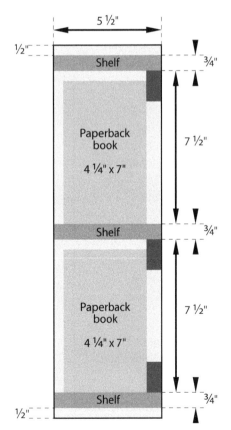

Figure 11-3. Dimensions for an end piece of the bare-minimum bookcase.

My plan is in Figure 11-3. Using one-by-six board, each shelf will be ¾" thick and about 5½" wide. I've added a ½" margin above the books. I added another ½" to the top and bottom of each end piece, so that it sticks out above the top shelf and below the bottom shelf, because I think this looks nice.

If you're inspecting my plan and wondering what the three small, dark-brown rectangles are, I'll get to them in a moment.

Now that I have a plan, I can add up the dimensions from top to bottom. Like this:

½" + ¾" + 7½" + ¾" + 7½" + ¾" + ½" = ?

Some people have problems with adding fractions. If you are one of those people, I suggest you convert the fractions so they all use the same units. For instance, ½ is the same as ²⁄₄, so the addition can be rewritten like this:

²⁄₄ + ¾ + 7 + ²⁄₄ + ¾ + 7 + ²⁄₄ + ¾ + ²⁄₄

Group all the fractions together, and group the whole numbers together, and you get:

¹⁷⁄₄ + 14

But ¹⁷⁄₄ is really 4¼ . So 4¼ + 14 = 18¼".

Alternatively, you can use decimals, and put the numbers into a calculator, like this:

0.5 + 0.75 + 7.5 + 0.75 + 7.5 + 0.75 + 0.5 = 18.25

You now have enough information to figure out how much wood you need. Three shelves, each 18"—that's a total of 54". Two vertical supports, each 18¼"—that's a total of 36½". Grand total: 90½". If you allow a little bit for the widths of saw cuts, and trimming the ends, and maybe avoiding a defect here and there, you may be able to get everything out of a single one-by-six board that is 96" long.

Figure 11-4 shows what I have in mind. The dark areas show possible places where you will cut out shelves and end pieces. Their actual locations will depend on your need to avoid any defects.

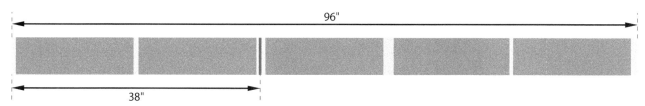

96"

38"

Figure 11-4. If you have to reduce the length of a 96" board to transport it home, the location of the red line is a suitable place to cut it. The exact positions of the sections that you remove from the board later will depend on any defects that you wish to avoid.

My main concern, here, is that if a 96" board is too long for you to get it home, don't let the lumber yard cut it in the middle. Ask them to make a cut 38" from one end, as shown by the red line in the diagram. That should be a fairly safe bet.

Making the Shelves

I suggest you begin in the usual way, by trimming your piece of lumber to make sure that the end is clean and square. You can suppress splintering with a sacrificial piece of two-by-four under it, and I think you'll need a guide piece over it, because your cut has to be precisely vertical. If the cut is at an angle, it won't look good when it butts up against the end piece.

You now have your reference cut, and you can measure your first shelf from it. You can use the sides of the shelf as they are, because a one-by-six is usually sold with pretty straight edges (so long as it has not warped), and in any case, the exact width of the shelves is not important in this project.

Figure 11-5. Measuring one 18" shelf.

Align one end of your 18" ruler with the cut that you just made, and place your speed-square against the other end of the ruler, across the wood, as in Figure 11-5. Hold the speed square in position while you remove the ruler and extend your pencil line across the board.

You will need to saw along the far side of this line, so allow enough space for the width of the saw blade, as in Figure 11-6.

Figure 11-6. Cutting the first shelf requires leaving room for the width of the saw.

Figure 11-7. Make sure the ends of the wood are precisely aligned.

Figure 11-8. Cutting along the edge of one shelf, to make a second shelf.

Now you have one shelf, you can use it as a guide, to make two more. First align the end of the shelf precisely with the end of your remaining length of board, as in Figure 11-7. You can place a flat object across them, such as the side of your speed square.

Clamp the two boards together, turn them around, and now the end of your shelf is your new guide for cutting another 18" piece, as in Figure 11-8. If the concept isn't clear, the diagram in Figure 11-9 may be helpful.

You can use this system to make multiple copies of a piece that you have cut, provided you do it the right way. Suppose you start with Piece A. You use Piece A to measure Piece B. Then you reuse Piece A to measure Piece C. Then you reuse Piece A to measure Piece D, and so on.

Don't use Piece A to measure Piece B, and then Piece B to measure Piece C, and then Piece C to measure Piece D . . . because if you make small errors in that sequence (perhaps because your cuts are not precisely vertical), the errors will add up, in the same way that the quality of a photocopy deteriorates if you make a copy of a copy of a copy.

After you cut the shelves, rounding their long front edges with sandpaper is a good idea. When you remove books or replace them, they tend to rub over the front edge of a shelf, and if that edge is not rounded, paint or polyurethane doesn't stick to it very well.

Reinforcing the Shelves

Going back to my list of requirements, the second one cited the tendency of store-bought shelves to sag when books are placed on them. The 18" shelf in this project is very short, and little paperbacks are

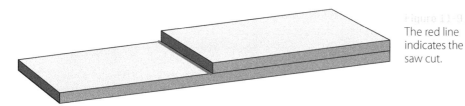

Figure 11-9. The red line indicates the saw cut.

relatively light, so you won't see it sagging. But just for future reference, how long can a shelf be, before it does start to sag noticeably?

Answering that question would require knowing what kind of wood will be involved, how much load it will carry, and how thick it is. However, if the length of a shelf increases, I can tell you how much more it will sag, relatively speaking. A bit of math gives the answer (you can skip this if you dislike math).

Suppose you double the length of a shelf, while the load per linear inch remains constant. The doubled length will cause the shelf to sag eight times as much. If you triple its length, it will sag 27 times as much.

This is because the sag is proportional with the cube of the length of the shelf. In other words, if L is the length, the sag is proportional to L x L x L.

But, there is some good news. If you double the *thickness* of the shelf, you *divide* the sag by 8. If you triple the thickness, you divide by 27, because, the sag is inversely proportional with the cube of thickness. So, if T is the thickness, the sag is proportional with 1/T x 1/T x 1/T.

Okay, that's the end of the math.

A Stiffening Strip

In practical terms, you can eliminate visible shelf sag by adding a strip to stiffen the shelf, which is what I want to do in this little bookcase. This will have two benefits.

First, you will be able to use the same design to make much longer shelves in the future, as shown in Figure 11-10.

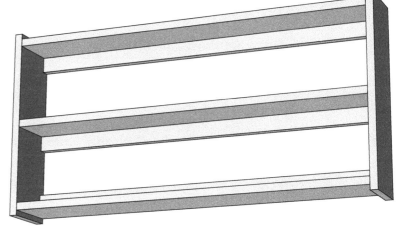

Figure 11-10. With a stiffening strip attached under (or above) each shelf, much longer shelves are possible without any visible sag.

The stiffening strip will be a piece of one-by-two hardwood, added under the back edge of each shelf. The hardwood can be oak, maple, or poplar, and I'm betting your local big-box store will have one of those. This will be stiffer, and will grip screws better, than comparable softwood. Incidentally, one-by-two is actually ¾" x 1½", but you probably guessed that.

The second advantage of these strips is that they are convenient for hanging the bookcase on the wall. If your wall has wooden studs in it, you find out where they are, drill holes in the stiffening strips at appropriate intervals, and use some screws that are 2¼" long. I'll explain what studs are, and the whole topic of hanging things on walls, in the very next project.

For the bottom shelf in this little test project with 18" shelves, a stiffening strip isn't necessary. The strips on the upper two shelves are sufficient to hang the bookcase on the wall, and the shelves are so short, sagging won't be a problem.

The two strips of one-by-two hardwood will be 18" long, to match your top and middle shelves. I think the easiest way to attach them to the shelves is by using screws, which will be unobtrusive at the backs of the shelves. You could use more pegs instead of screws, but that would take longer.

Figure 11-11 shows a plan for drilling screw holes in the two shelves. All measurements derive from the top-left corner, because if you make a series of small measurements from one hole to the next, cumulative errors can occur.

Figure 11-11. Where to locate the screw holes in your shelves. Countersink each hole.

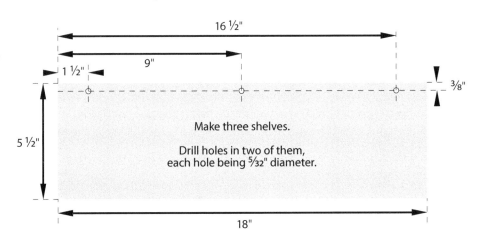

Make three shelves.

Drill holes in two of them, each hole being ⁵⁄₃₂" diameter.

Figure 11-12. Make a pilot hole in the piece of one-by-two hardwood by drilling through the screw hole in the one-by-six board.

After you drill the ⁵⁄₃₂" holes in the shelves, countersink each one to accept a flat-headed wood screw. Your next step is to clamp one of the shelves to one of the stiffening strips, in correct alignment. You can then use a ⅛" bit to make pilot holes in the stiffening strip for your #8 screws. (Check back to Figure 10-34, and you'll see that if you are inserting a #8 screw into hardwood, a ⅛" pilot hole is appropriate.)

Simply drill with your ⅛" bit through each screw hole, into the strip below, as shown in Figure 11-12. This way, the pilot holes are automatically aligned. Note that I used a piece of scrap two-by-four in the background, to keep the shelf level during this process.

Insert the screws, without removing the clamps until the sections of wood are screwed tightly together. Then repeat the process for the second shelf.

Making the End Pieces

Figure 11-13 shows where you will be drilling holes for the pegs in the end pieces. The darker strips of color in the figure are included just to remind you where the shelves will be. All the measurements are made from one end of the wood, because if you make a sequence of short measurements, one after another, you will tend to get cumulative errors.

Remember, these holes are for the pegs, and *do not go all the way through the wood.* You need to put a piece of masking tape around your drill bit, ⅝" from the tip of the bit, to remind yourself when to stop drilling.

A ¼" bit will be appropriate for the pegs, but I suggest you begin with a ⅛" bit, which will be easier to control. This is shown in Figure 11-14. Try to make the holes vertical, so that when you enlarge the holes and put pegs into them, the pegs will not lean at an angle.

Once you have the holes correctly located, you can enlarge them to ¼". Incidentally, you don't need to worry if there is just a little bit of splintering around the edges, because the ends of the shelves should conceal this. The full-size holes are shown in Figure 11-15.

Before you continue, make sure there are no wood chips in the holes. Blow into them, using a drinking straw and closing your eyes if you want to protect yourself from dust flying into your face.

Now dribble some carpenter's glue into each of the holes, smear some more glue around each of the pegs, and push the pegs firmly into the holes. Tap them

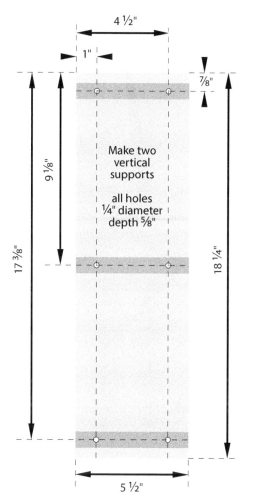

Make two vertical supports

all holes ¼" diameter depth ⅝"

4 ½"

1"

⅞"

9 ⅛"

17 ⅜"

18 ¼"

5 ½"

Figure 11-13. Locations of holes in each end piece for the bookcase.

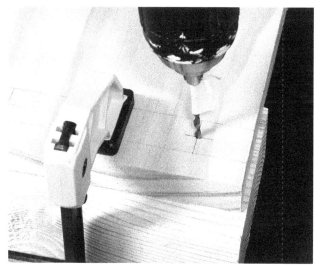

Figure 11-14. Begin with a ⅛" bit, and don't go deeper than ⅝".

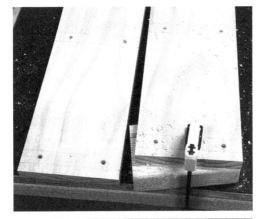

Figure 11-15. Plugs can be glued in place, as soon as debris has been removed from the holes.

Figure 11-16. All the plugs have been glued into the end pieces.

Figure 11-17. Mark the positions of the plugs on the matching shelf.

Figure 11-18. Extend the marks around to the edge of the shelf, using your speed square.

lightly with a hammer, if necessary, or rotate the peg while you continue to push it, to make sure it is completely seated. If you see some glue around the edge of the hole, wipe it away promptly with a wet rag or paper towel. The result should look like Figure 11-16.

After your gluing is done, you need to make holes in the ends of the shelves that will exactly match the positions of the pegs. Of course the pegs should all be the same distance from the edges, but little errors may have crept in. So, you need to deal with the pegs where they actually are, not where they should be.

Number the backs of the shelves in pencil, so you won't get them mixed up. Then stand each shelf beside the pegs that must fit into it, and draw around the pegs, as shown in Figure 11-17. Keep the pencil straight, so that the lines are accurately transcribed.

After you have penciled the outline of each peg, turn the shelf and extend the pencil marks down its edge, as in Figure 11-18. Make a mark where the peg will be centered between these lines, an equal distance from the top and bottom of the shelf.

Now you need to drill holes in the ends of the shelf. I think this will be easiest if the shelf is vertical. One way to secure the shelf vertically is by clamping it to a block of two-by-four at the bottom, and then clamping the two-by-four to your work surface. This is visible in Figure 11-19, which also shows a preliminary hole being drilled with a ⅛" bit. Some people prefer to clamp the shelf flat on the work area and drill into it horizontally, but doing it vertically is easier for me.

When all your drilling is done, you're ready for the final assembly. Clear your work area, and clean any dust or fragments off the parts of your project, especially the ends of the shelves. Remove little splinters from around the holes that could interfere with the shelves fitting tightly.

You will need to clamp the whole bookcase together, while the glue is drying, but your clamps aren't long enough to reach across it. How can you address this problem? My suggestion is shown in Figure 11-20, using a couple of loops of nylon rope around blocks that spread the load.

But first, you need to do a rehearsal. Without applying any glue, make sure that everything fits together. If some of your holes are in the wrong place, you need to discover this now, not after the glue has been applied. Holes can be enlarged with your drill, or additional holes can be made if a measurement error resulted in holes being in the wrong place. An enlarged hole will make a weaker joint, but so long as some of the peg makes firm contact, it may be acceptable.

If everything looks okay, disassemble your bookcase, squirt glue into each hole, smear more glue on each peg, put everything back together, and tighten your clamps in the loops of rope.

Use your speed square to make sure that the angles between the parts of your bookcase are 90 degrees. If necessary, you may have to apply some sideways force to square it up. This can be done with another piece of rope or string, tied around a leg of your workbench or table.

Figure 11-19
The end of the shelf is drilled while the shelf is clamped vertically.

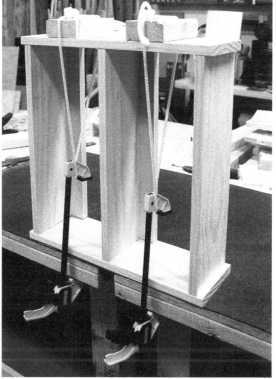

Figure 11-20
An improvised system for applying clamps that aren't long enough to reach across the bookcase.

My finished bookcase is shown in Figure 11-21, after being coated with polyurethane. In the next project, I'll show you how to hang it (and other objects) securely on a wall.

Other Ideas

Pegs can be used to join boards edge-to-edge. In fact, that's how furniture was made before the widespread use of composite materials such as plywood.

Ideally you should use a planer to put an absolutely straight edge on each piece of wood before you join it. A planer is a heavy expense, however, and if you just choose boards that are reasonably free from imperfections, an edge joint can be hard to see.

Using edge-to-edge joining, you can make traditionally styled furniture that avoids the messy edges of plywood. I constructed the little nightstand in Figure 11-22 from one-by-six maple boards joined edge-to-edge (with the exception of the back, which is ³⁄₁₆" plywood). The grain patterns in the top, the sides, the door, and the interior shelf were chosen to conceal the joints. The project is shown before a finish was applied. Fabrication took about 25 hours. Power tools were used extensively.

Figure 11-21. The finished bookcase, coated with polyurethane.

Figure 11-22. A fabrication project made entirely from boards joined edge-to-edge with wooden plugs (except for the plywood back).

Hanging On the Wall

While this is not a home improvement book, one tool-related task in the home is so common, so necessary, and so often misunderstood, I think it should be included here. I'm going to show you how to hang a small object on a typical wall, and after that I'll suggest how to hang larger objects on walls that are not so typical. At the end, I'll deal with the general topic of metal and plastic accessory items such as anchors, brackets, and steel wire.

A Bookcase on the Wall

In the United States, most homes are built around a framework of two-by-four or two-by-six lumber, such as the example shown in Figure 12-1. Some multistory buildings are similarly framed, but large apartment buildings of brick or concrete are more likely to have internal partitions framed with thin steel. I'll get to them a bit later. Initially, I will assume that your home is wood-framed.

NEW TOPICS IN THIS CHAPTER

- How to hang things up that won't fall down
- Steel and plastic accessories

YOU WILL NEED

- Tape measure (optional)
- Stud finder (optional, see text for explanation)
- Level (optional)
- Sewing needle (carpet needle preferred)
- Deck screws, 2¼", quantity 3

Also, as listed previously: Ruler, work gloves, pliers, electric drill and drill bits, countersink, awl, masking tape, screwdriver or electric screwdriver, and safety glasses (optional).

Check the Buying Guide on page 248 for information about buying these items.

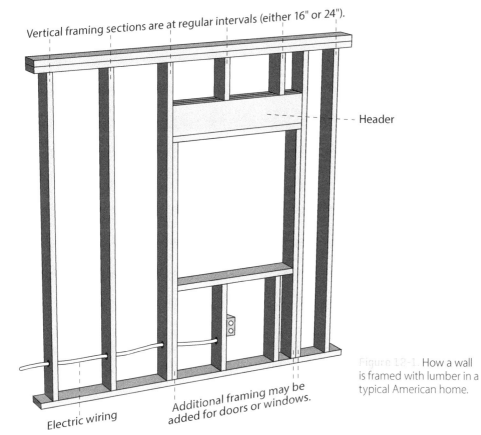

Vertical framing sections are at regular intervals (either 16" or 24").

Header

Electric wiring

Additional framing may be added for doors or windows.

Figure 12-1. How a wall is framed with lumber in a typical American home.

After the vertical pieces of lumber are installed in a wall, they are known as **studs**. They are spaced either 16" apart or 24" apart; the wider spacing saves a little money, but is less common.

If you are wondering where the numbers 16 and 24 come from, it's because 3 x 16 = 48, and 2 x 24 = 48. Standard sheets of **drywall** are 48" wide, enabling its edges to be nailed or screwed to the framing.

Drywall is also known as **plasterboard**, **wallboard**, **Sheetrock**, and **gypsum board** (Sheetrock is a trademark.) It consists of compressed plaster, or gypsum, often sealed between two sheets of paper and sold in sheets ½" or ⅝" thick. The advantages of drywall are that it is speedy to install, fire-resistant, and relatively easy to finish, especially if a texture is sprayed onto it.

The big problem with it from a homeowner's point of view is that it has very little structural strength. When you poke a hole in drywall, powder comes trickling out. If you bang a nail into it, the drywall does not hold the nail securely. When you drive a screw into it, a small amount of pulling force will cause the screw to fall out. There are ways to deal with these issues, but the best strategy is to avoid dealing with drywall altogether, by finding a wooden stud inside the wall and driving a screw through the drywall, into the stud.

For this purpose, you need some pliers, a needle, and (optionally) a gadget known as a **stud finder**. I'll begin by describing the procedure if you don't own a stud finder.

Stud Hunting

Using trial and error, you can get a rough idea of where a stud is located by knocking with your knuckles on the wall and finding the areas which don't sound so hollow. Studs are likely to be located in those areas. You may also find studs by looking for imperfections where sheets of drywall were joined together. Another clue is that an electrical outlet is usually attached to the side of a stud inside the wall.

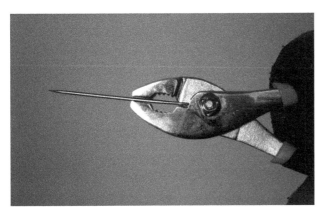

If you have a rough idea where a stud is located, you can find it more precisely by grasping a needle in a pair of pliers and poking it into the wall at horizontal intervals of 1½". You have to push hard to get the needle through the drywall, but after it has penetrated, if there is no stud in that location, the resistance to your pushing will suddenly disappear. On the other hand, when you do find a stud, the needle will bump into it, and won't proceed any further.

A carpet needle is best for this purpose, because it's longer and less likely to bend than a regular sewing needle. It can be gripped with the pliers as shown in Figure 12-2.

Figure 12-2. Gripping a carpet needle in some generic slip-joint pliers.

If you make the holes in an inconspicuous area (perhaps down near the baseboard), they may be so small, no one will notice. Alternatively, at the end of the job you can fill them with tiny amounts of **caulking**. This is mostly sold in large tubes that fit a **caulking gun**, but you can also buy it in small, squeezable plastic tubes.

Once your needle has hit a stud, you need to make a couple more holes at smaller intervals to find the edges of the stud. When you have the edges, you can make a pencil mark half-way between them, which will be the center of the stud.

If you tie a heavy object, such as your steel ruler, to the end of a piece of string, the string will hang vertically. You can hold the top end at a level where you want to put a screw into the wall, and move it along till the bottom of the string coincides with your mark for the center of the stud. The string now coincides with the stud from top to bottom.

If you need to secure your bookcase (or other object) at more than one point, measure 16" horizontally from your first stud, and see if another one is there. If not, move along another 8" to see if your home is in the minority framed with studs at 24" intervals.

What Can Go Wrong

If you simply cannot find a second stud, the first one may have been an "extra." Check back to Figure 12-1, and you'll see that extra studs are added around windows. They may also be added around doors, or where one wall intersects with another.

Suppose the first stud that you find is on the right-hand side of the window in the illustration. Measure 16" farther along from that, and you won't find anything.

Another possibility is that you may stick your needle where doubled studs are sandwiched together, and the needle will just happen to slide between them. Doubled studs are often used beside windows and doors. Again, if you start in a blank area of the wall, this is unlikely.

What if your needle will only penetrate a tiny amount, such as 1/16"? Most likely you just happened to hit the head of a screw or nail that is hidden below the surface of the paint. Move up or down 1/2", and that problem should be solved.

Generally speaking, building codes require that wiring must be strung through holes that are centered in the studs. Your needle should not be long enough to reach that far. In cases where wiring or a gas line is located closer to the surface of the wall, it will be protected with a metal plate on the side of the stud, so that you can't drive a screw into it accidentally. It is possible, but unlikely, that you may hit a metal plate with your needle, in which case, once again, you will have to go a little higher or lower on the wall.

Making It Level

I will assume that you found a first stud, and marked a point where you will drive in a screw, and now you have found a second stud, in which to drive a second screw. You need to make sure that your screws will both be the same height on the wall.

For this purpose, you can place a **level** between them, and nudge its end up or down until it's horizontal. A level is a straight bar of aluminum, wood, or plastic containing at least one small curved tube full of fluid with a bubble in it. When the level is horizontal, the bubble is centered between two marks. Figure 12-3 shows a basic level, while Figure 12-4 shows a version that contains a laser pointer.

Figure 12-3. This basic level can be used to verify vertical, horizontal, or 45-degree angles .

Figure 12-4. This level incorporates a laser pointer.

If you prefer not to buy a level, you have an easy alternative: measure upward to make sure both your anchor points in the wall are the same distance from the ceiling. This is more reliable than measuring downward to the floor, because floor coverings introduce some uncertainty. You can use a tape measure for this purpose, or reuse your piece of string.

The string method is fairly reliable, but just to make sure, you can stretch a piece of masking tape between your two chosen anchor points, step back, and see if the tape looks level.

After verifying your anchor points, push an awl into each one of them and wiggle it around a bit to make a hole big enough for a screw to go in. The screw won't bite into the drywall, so you may as well enable it to get straight to the stud.

Assuming your studs are 16" apart, you can go back to your bookcase and drill two 3⁄16" holes 16" apart in the stiffening strip for the center shelf. This should be no problem, as I designed the shelves to be 18" long (and now, you know why). Therefore, each screw hole should be 1" from the end of the shelf. Position the screw holes 1⁄2" from the lower edge of the stiffening strip, because this will be easier to reach with your screwdriver when you are attaching the shelf to the wall.

After you drill each hole, countersink it to accept a flat screw head. Figure 12-5 shows a 2¼" screw that has been pushed through a hole in the inner side of the stiffening strip, and is ready to be inserted into the wall. Figure 12-6 shows the screw being driven in. A long screwdriver is more convenient for this than a short one, as it keeps your knuckles away from the shelves.

Figure 12-5. A hole has been drilled in a stiffening strip under the center shelf in the bookcase from Chapter 11. A 2¼" deck screw has been pushed through the hole.

Drilling a pilot hole into the stud is usually unnecessary because (a) the stud is soft pine, (b) only about ¾" of the screw will reach the stud after passing through the stiffening strip of the bookcase and the layer of drywall, and (c) the thickness of the drywall would prevent a small drill bit from penetrating the stud very deeply anyway.

What if your studs are 24" apart instead of 18" apart? In that case, you can secure your bookcase with two holes vertically above each other, both sharing one stud. Each screw will go through a separate stiffening strip in a separate shelf. This will not be quite as secure as the two screws placed horizontally at opposite ends of one shelf, but for a bookcase as small as this, it doesn't matter.

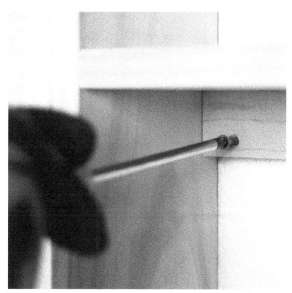

Figure 12-6. Driving the screw into the stud inside the wall.

My bookcase is shown attached to a wall, with a full load of books and ornaments, in Figure 12-7.

Figure 12-7. The bookcase, with books and ornaments, on the wall.

The stud finder shows an initial reading when it is about 4″ from the stud inside the wall.

The finder moves closer to the stud.

The finder is now about ¾" away.

The finder has located the edge of the stud, with an accuracy of about plus-or-minus ½".

Better Stud Finding

What about the gadget that I mentioned earlier—the stud finder? Many versions are available, the simplest type being a little handheld, battery-powered device that senses the change in electrical capacitance between a hollow section of the wall and the location of a stud. It will have an LED or some other method to tell you when it thinks it has found the edge of a stud.

Modern scanning devices are more sophisticated, easier to use, and able to detect objects such as wires and pipes inside walls. They are more expensive, however.

To use a basic stud finder, you have to initialize it by pressing a button while you hold it against the wall in a section where there are no studs. Then drag the sensor horizontally until it indicates that it has found the edge of a stud. In the finder that I use, the sequence of events is shown in figures 12-8, 12-9, 12-10, and 12-11.

Having located one edge of the stud, you can find the other edge by repeating the process from the opposite direction. The finder is not entirely accurate, so you'll still need the pliers-and-needle system to make sure.

There is one other stud-finding option that I know of. You can poke a hole into a hollow part of the wall with an awl, then bend a metal coat hanger into an L shape, thread it through the hole, and rotate it till the end inside hits a stud. Extract the coat hanger while trying not to change its angular orientation, and the end of it should show you where the stud is.

The only problems with this method are that it does require you to make a hole in the wall, and if the wall contains insulation, it will get in the way.

Alternatives

Perhaps you are unable to hang something from a stud in your home, either because the studs are in the wrong place, or because you have walls that were not constructed with wooden studs. You may have been wondering when I will get around to your situation. I am now happy to do so.

Here are some possibilities that you may encounter:

- Plaster over brick, usually in old buildings.
- Thick plaster over wooden **laths**, in very old buildings. (Laths are thin horizontal strips of wood, nailed to wooden studs. The laths are just adequate to support plaster that is applied to them by hand, while it is wet.)
- Plaster over cinder blocks or concrete.
- Drywall with wooden studs, but you want to hang something in a location where there are no studs.
- Drywall attached to steel studs.
- Drywall on furring strips over cinder blocks.
- Alternative and experimental materials, in recent buildings.

Incidentally, **cinder blocks** are made primarily of concrete mixed with sand, fine gravel, or industrial wastes. They are properly known as **concrete masonry units**, or **CMU**s. Various types can be referred to as **hollow blocks**, **concrete blocks**, **cement blocks**, and **breeze blocks**. A cinder block is shown in Figure 12-12.

Figure 12-12 A cinder block.

I'll start with the type of wood-framed wall I have already dealt with. What if you absolutely, positively have to hang something in a location where there isn't a stud?

Empty Sections of Drywall

For lightweight loads, you can use an appropriate **drywall anchor**. The simplest, cheapest, and most common type is also the least effective: a hollow plastic plug, sometimes known as a **fluted plastic plug**, and sometimes as a **Bantam** plug (which is a trademark).

If you buy an object such as a smoke detector, or a small mirror designed to hang on the wall, it may be supplied with a couple of drywall anchors of this type. You make a small hole in the drywall, then insert the plug, and then drive a screw into the plug, causing the plug to expand. Because of the small size of the plug and the lack of structural strength in drywall, this kind of fixing device is not very secure.

A more effective but more expensive plastic anchor is shown in Figure 12-13, in two sizes. These may hold heavier loads, perhaps up to 20lbs, so long as the load is applied at 90 degrees to the anchor and does not pull outward from the wall.

First, you use an awl to poke a hole in the drywall about ¼" in diameter. Then you insert a large Phillips screwdriver in the cross-shaped socket at the end of the anchor, and screw it into the hole. The threads on the anchor are so wide, they make a fairly good connection with the powdery stuff in the drywall.

You can now insert a regular Phillips screw into the hollow core of the anchor. The screw should be the one supplied with the anchor, but if you need a longer one, it should be about the same diameter as the one that was supplied.

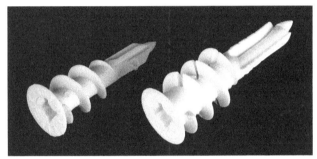

Figure 12-13. Two sizes of plastic anchors for drywall.

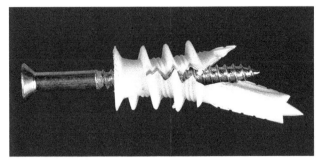

Figure 12-14. What happens when you insert a screw into a plastic anchor.

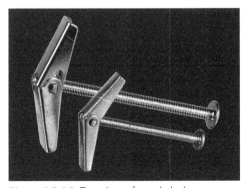

Figure 12-15. Two sizes of toggle bolt.

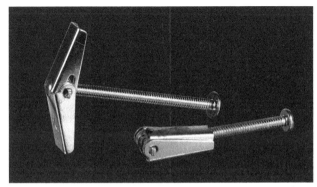

Figure 12-16. A toggle bolt with its toggles closed, and another with its toggles open.

The screw not only makes a fairly secure connection with the plastic, but also causes the anchor to expand, and eventually breaks it open, as shown in Figure 12-14. This actually improves the security of the anchor in the drywall by making it less likely to pull out.

Naturally you have to choose the length of your screw appropriately. If you are screwing through a piece of ¾" wood, as in the bookcase project, your screw needs to be around 1½" long to go through the wood and deep enough into the anchor.

Careful measurement will be required to match the location of the anchor with the position of the screw— or you can make a mark through a hole in the object that you are hanging, and onto the wall, then insert an anchor there.

Another option for hollow drywall is a **toggle bolt**, two sizes of which are shown in Figure 12-15. The toggles of the bolt are spring-loaded, so that they flip open as shown in this figure. You hold the toggles closed, with finger and thumb, to insert the bolt through a hole in the drywall. When the bolt is through the hole, the toggles open against the inside of the wall, and you can tighten the bolt against them. Figure 12-16 shows one toggle bolt open, and another closed.

The toggles require a relatively large opening in the drywall. You can create this with an awl, just by pushing it in and out, rubbing it against the sides of the hole. The friction causes the shaft of the awl to eat into the drywall. You'll need to place a waste bin below, to catch the dust.

A disadvantage of a toggle bolt is that the bolt supplied with it may not be long enough. In Figure 12-16, the toggles are about ¾" long and must push all the way through drywall before they can open. If the drywall is ½" thick, and you want to hang something such as the bookcase where the strip of wood is ¾" thick, you may think that a 2" bolt will be just sufficient, but actually it has to be a fraction longer, because the end of it must screw into a threaded collar mounted between the toggles. Personally I have often substituted bolts that are 2½" or 3" long. These have to be purchased separately.

Drywall Mounting With Steel Studs

A multistory building constructed with brick or concrete is likely to use **steel studs** inside walls that subdivide the interior space. A steel stud has a cross-section like a flattened U, as if someone wrapped thin metal around three sides of a two-by-four.

A stud finder will find a steel stud just as easily as a wooden stud, but once you find it, what are you going to do with it?

Fine-threaded drywall screws have sharp points designed to penetrate the steel, but these screws may not be long enough if you want to hang something substantial on the wall. Consequently, you may have to drill a pilot hole and use a regular sheet-metal screw.

First make a hole in the drywall with your awl, until you hit the stud. Clean as much plaster dust out of the way as you can, because even a small amount of the white powdery stuff will blunt a regular drill bit. Once you have a hole, if you shine a flashlight into it, you should see the steel stud gleaming in there.

Press hard while using a drill bit of around $\frac{3}{32}$" diameter at a slow speed, and it should make a hole in the thin steel fairly easily. Don't allow the drill to penetrate deeper, after it makes the initial hole. Now you can use a #8 sheet-metal screw to engage in the hole. Be gentle, because if you turn the screw with too much force, it may pull out of the stud.

This last sentence tells you that while steel studs can support a heavy vertical load, they are very poor at resisting a pulling force. If you are hanging something such as a kitchen cabinet that will contain a lot of heavy dishes, you may have a problem, because the greater depth of the cabinet will enable it to apply a strong tipping force. I would be cautious about using steel studs for that job.

Working With Masonry

Suppose you have a wall made of brick or cinder blocks, which are either exposed or covered directly with plaster. This is generically described as a **masonry** wall. It should be able to support as much weight as you choose to hang on it. Better still, you can use any part of it. Your only problem is that you cannot drive ordinary screws into it.

What you have to do is drill a hole in it using a special **carbide-tipped** masonry bit. Ideally you want your hole to go into a brick or a cinder block, not the layer of mortar between them, as mortar is not as strong.

A battery-powered drill is probably powerful enough for a ¼" bit of this type, although you may have to lean on it for a while, depending on what kind of masonry you are dealing with. Even in a brick wall, you may be surprised to find that some bricks are much harder than others.

After you drill a hole, you can insert a **masonry anchor** that consists of a cylindrical plug made of plastic, lead, or wood. The anchor has a hole in the middle, so that you can drive in a screw. The screw forces the anchor to expand, making a tight fit in the hole. Friction keeps the anchor from coming out.

If you are attaching a strip of wood to the wall, you insert the screws in the holes in the wood, which must line up precisely with the anchors. Peek around the edges of the wood, using a flashlight if necessary, to make sure that each screw will go into the center of an anchor. You do not want the screw to go between the anchor and the edge of the hole that it's in.

You have to use a screw that is large enough to make the anchor expand sufficiently. Typically a #10 screw is good with a ¼" lead anchor, but in hard brick which resists the expansion of the anchor, you may need a #8 screw. Don't use excessive force to insert a screw; it can fracture the anchor inside the hole. The best approach is to lubricate the screw thread with some oil.

Sleeve bolts are an alternative to masonry anchors. A bolt is mounted in a metal sleeve for insertion into a hole in the wall. When the bolt is tightened, the sleeve expands.

A **sleeve nail**, also known as a **nail drive anchor**, is a nail in a metal sleeve that expands when the nail is driven in; this type of masonry anchor is more difficult to remove.

Blue concrete screws are yet another alternative for masonry. They are sometimes known by their brand name, **Tapcon**. Because their hardened threads can cut into concrete, brick, or cinder blocks, no anchor or sleeve is necessary. You simply drill a pilot hole that is precisely the right size for the screw.

Drilling holes in masonry is not much fun, but being able to put shelves wherever you want them is a nice capability to have.

Drywall Over Masonry

When the interior of a building is renovated, drywall may be added in front of masonry. This often happens in a basement, when the home owner decides to cover some naked cinder blocks. A common strategy is to attach **furring strips** to the masonry before screwing sheets of drywall to the strips. An example is shown in Figure 12-17.

This creates a difficult situation when you want to hang something on the wall. A furring strip is typically about 1" x 4". In other words, the wood behind the drywall is only 1" thick. If you want to drive a screw into it, the screw will have to be exactly the right length. If you want to use a toggle bolt in the drywall between the furring strips, the gap may not be large enough for the toggles to open. A plastic drywall anchor may be your only easy option. You should also be cautious in case someone has run plumbing or wiring in the very thin space between the furring strips.

Figure 12-17. Furring strips added to a wall of cinder blocks. Sheets of drywall will be attached to the strips.

To hang something heavy, you may have to cut a rectangular opening in the drywall with a utility knife, to get access to the cinder blocks or brick behind it. Insert a wooden block in the rectangular opening, and secure the block with masonry anchors. Then you can hang a heavy weight from the block.

Old-School Plaster

In an old house with walls of plaster on top of laths, you can search for studs behind the laths, but they may be difficult to find. Another option is to treat the plaster as if it is masonry, and use masonry anchors. Back in the day, plaster was often applied in a very thick, heavy layer that had quite a lot of structural strength. You'll have to use trial and error to find out exactly what you are dealing with. Even heavy-duty plaster may crack if you apply too much compression to it from an anchor.

Difficult Objects

Some objects are difficult to hang on any kind of wall. I deliberately designed the bookcase in Chapter 11 so that it would be easy, with horizontal strips of one-by-two hardwood. But what if you want to hang a store-bought bookcase (for example) that doesn't have such convenient attachment points?

One option is to put a wooden strip under the shelves, in the same way as the stiffening strips in the basic bookcase. The disadvantage of this idea is that if you have multiple shelves, you need multiple strips.

Alternatively, you can just put one strip under the top shelf and suspend the bookcase from that. But this means that all the weight of the lower shelves will be transmitted through the end pieces, and then through the pegs to the top shelf. Store-bought shelves are not designed for this. The end pieces may be fabricated from thin particle board that can fail unexpectedly. The pegs in them can easily pull out.

If you are working with shelves of this type, the simplest option is to mount a separate horizontal strip of 1" x 4" or 1" x 3" wood on the wall, equal in length to the width of the bookcase. This strip will take the weight of the bookcase standing on top of it.

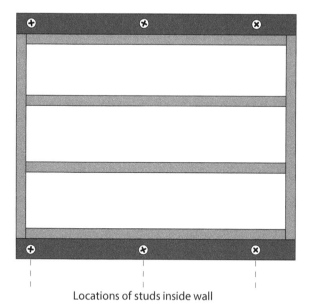

Locations of studs inside wall

Figure 12-18. When a long wooden strip is used to support a bookcase, the screws can be aligned with studs in the wall, regardless of the exact location of the bookcase.

Attach the wooden strip to the wall by screwing it into wooden studs (if it's that type of wall) or using masonry anchors (if it's a masonry wall). Plastic sheetrock anchors are not strong enough to support the weight of a medium-sized or large-sized bookcase, but they're not necessary anyway, as the screws in the block can be placed wherever the studs are inside the wall, especially if you plan to conceal them with filler and then add a coat of paint. The screws do not have to be arranged symmetrically with the bookcase. Figure 12-18 shows what I mean.

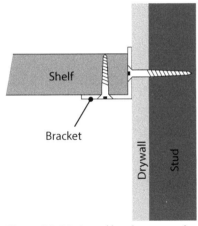

Figure 12-19. A steel bracket to attach a shelf in a bookcase to a wall.

A second strip of wood is used to secure the top of the bookcase, to prevent it from falling forward. Screws will be driven down through the bottom shelf into the bottom strip, and up through the top shelf into the top strip.

The best I can say about this system is that it is secure and doesn't require alignment with studs in the wall. It's not very elegant, though.

A less obtrusive alternative is to use **small steel brackets** instead of wooden strips. Figure 12-19 shows a cross-section of this arrangement. First the brackets are screwed onto the wall, at least two for each shelf, and spaced vertically by a distance equal to the spacing of the shelves. Then the shelves are hung on the brackets. If the shelves are made of particle board, it may be a good idea to use nuts and bolts to clamp them to the brackets, instead of wood screws—especially if the shelves are thin. I'll be getting to the whole topic of nuts and bolts in Chapter 14.

A good general rule for hanging anything on a wall is that it should be strong enough to take twice the load that you expect. Bear in mind that people often end up piling much more stuff on shelves than anyone originally expected.

Other Uses for Small Brackets

Now that I have mentioned steel brackets, I should show some examples and suggest some applications. Figure 12-20 shows a selection of brackets that you can find cheaply in almost any hardware store. The cutting mat that they are standing on is divided into 1" squares. The item in the foreground is a **mending plate**.

Figure 12-20. Some cheaply available steel brackets.

Figure 12-21 shows a variety of larger brackets.

Suppose you are building a large box that is going to carry a heavy load, and you want to be sure that the 90-degree joint between two sides of a box is really secure. You can reinforce it with internal steel brackets. This is definitely not an elegant solution, but sometimes a heavy-duty application requires you to use unattractive hardware. To hold the brackets in place, bolts can be used. Screws might pull out of the wooden sides of the box.

If you're familiar with speaker cabinets designed to be transported to rock concerts, you know that they have heavy

Figure 12-21. Larger bracket and mending plate.

molded plastic protectors around each corner. These protectors help to hold the cabinet together, in addition to protecting it from impacts. They are like three-dimensional brackets.

When you have furniture that stands on the floor, such as a free-standing wardrobe, a hidden bracket can hold it against the wall, to prevent it from falling forward. This may be necessary if the furniture is tall, the floor slopes, or you live in an earthquake area.

Brackets can also be used to repair old furniture that is broken. When a glue joint dries out and pulls apart, a hidden bracket can hold it together. You may also want to use a mending plate, which is just a flat strip of metal about 2" long with two or more holes in it.

Figure 12-22. If this kind of shelf support is used, the topmost screw must be as long and large as possible, making secure contact with a wooden stud inside the wall.

Shelf-Support Brackets

How about the really large, industrial-style steel brackets that are specifically designed to support shelves? The type that I like is shown in Figure 12-22. These are much stronger than the type that is usually stamped out of thin steel and painted white, gray, or black.

You have to be aware of the force created by a load on this type of bracket. When weight is added, it exerts a turning force that tries to pull the top screw out of the wall. To support a heavy load such as that shown in Figure 12-22, the top of the bracket should be secured by a heavy carriage screw.

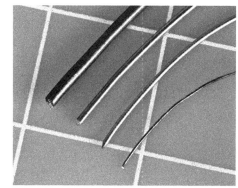

Figure 12-23. Galvanized wire is available in many sizes, and is useful for many tasks other than hanging pictures.

Other Metal Parts

Some people who are serious about carpentry tend to be purists who dislike mixing any metal parts with wood. I can share that outlook when I'm making something that is supposed to be beautiful, but when I have a purely practical objective, I just want to achieve it as simply and as effectively as possible.

Consequently, in my workshop, I have a lot of steel parts. There are various thicknesses of **galvanized wire**, as shown in Figure 12-23. I have a variety of **hose clips**, which are useful for many tasks other than clamping hoses (see Figure 12-24). And many sizes of **screw eyes** and

Figure 12-24. Hose clips are a quick and easy way to clamp things together.

Figure 12-25. Eye bolts (on the left) and screw eyes (on the right) are useful if you ever want to hang something heavy from an overhead beam.

Figure 12-26. Your local hardware store will have a fascinating variety of hinges, if you dig around in their metal-accessory section.

Figure 12-27. The paired items on the left can secure a cupboard door. The bolt and the hook-and-eye set are very traditional ways of doing the same thing, with a little more security.

eye bolts, shown in Figure 12-25. If you look at these items, can you start to imagine ways to use them?

I don't need any any of this stuff right now, but in the future, if I want to hang a bicycle from the rafters, or suspend a heavy lamp from the ceiling, or reattach a kitchen cabinet that has pulled away from the wall, a large and varied stock of hardware can be useful.

In addition, steel parts are indispensable for some applications. Hinges, for instance (see Figure 12-26). They come in many different sizes and styles. And then there are little devices to secure a cupboard door, as shown in Figure 12-27.

Wood is very versatile and can be lovely to look at, but sometimes a little hidden help from steel components makes a task easier.

Chapter 13
A Monster Truck

A tenon saw can only cut in straight lines, but many tools exist to cut around a curve, and some of them are inexpensive (see the Holes and Curves Fact Sheet on page 164 for details). For this project I want you to use one of the most basic curve cutters of all: a **coping saw**, pictured in Figure 13-1.

NEW TOPICS IN THIS CHAPTER

- Saws that cut curves
- How to enlarge and shape a hole

See next page for a materials list.

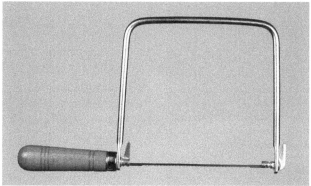

Figure 13-1. A coping saw.

The goal of this project is to make a simple toy for a very young child. Maybe you have a young relative, or the son or granddaughter of a friend, who will enjoy something that has wheels, no sharp corners, and can be thrown around without falling apart.

I'm thinking of a toy truck. But not just any toy truck. A monster truck, because that should be more interesting.

Getting Acquainted with a Coping Saw

The first thing you may notice about a coping saw is that the teeth of its skinny little blade seem to be the wrong way around. See the closeup in Figure 13-2.

Figure 13-2. The teeth of a coping saw are oriented to cut on the pull stroke, not on the push stroke.

In fact, a coping saw is designed to cut when you pull it, not when you push it. This is because the blade is thin and has no stiffness. It is maintained in a state of tension by the springy steel frame of the saw. When you push the saw, you tend to reduce the tension in the blade, while when you pull the saw, you increase the tension in the blade.

YOU WILL NEED

- Two-by-four pine, any condition, length 9"
- One-by-six pine, a few knots, not warped, length 18"
- Square dowel, hardwood preferred, ¾" x ¾", length 6"
- Round dowel, hardwood, ⅜", length 12"
- Round dowel, hardwood, ¾", length 6"
- Coping saw and spare blades
- Metal files (flat, round, and half-round)
- Cardboard, any thickness, size 3" x 5"
- Finishing nails, 1¼"
- Wood screws, #8 x 1½", flat head, Phillips, quantity 5

Also, as listed previously: Tenon saw, miter box, trigger clamps, ruler, speed square, rubber sanding block, work gloves, awl, masking tape, hammer, pliers, utility knife, electric drill and drill bits, countersink, screwdriver or electric screwdriver, dust mask (optional), safety glasses (optional), utility saw (optional), plywood work surface, sandpaper, cardboard size 12" x 12", and epoxy glue and hardener.

Check the Buying Guide on page 248 for information about buying these items.

Some people turn the blade around so they can use the coping saw in "push mode," but this does entail the risk of the blade popping out if you push too hard.

The blade is easily removed, not just so that you can reverse it or replace it, but so that you can thread it through a hole that you have drilled in a piece of wood. While the blade is in the hole, you can reinstall it in the frame of the saw. Now you can saw around the inside of the hole to change its shape.

You'll need to know the procedure for removing the blade, so let's deal with that now. Its teeth are sharp, and you may want to wear gloves.

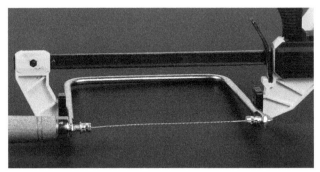

Figure 13-3. If you unscrew the handle of the saw by about ½", then clamp the frame, the blade is released from its mounts.

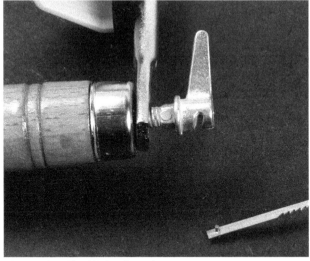

Figure 13-4. The blade released.

The wooden handle of the saw is mounted on a screw thread. When you turn it counterclockwise, it gradually releases the tension in the blade. If you unscrew it far enough, eventually the blade comes off completely— but now you have a loose handle, as well as a loose blade. I prefer to unscrew the handle just ½", then grab the frame of the saw in a clamp, as shown in Figure 13-3. Squeeze the clamp a little, and the blade should drop out of its mounts, as in Figure 13-4. You may need to practice this a couple of times, to get it right.

After you replace the blade in its mounts, retighten the handle. This pulls in the frame of the saw. Tighten the handle of the saw completely before you cut wood, and make sure that the metal tabs at each end of the blade are both pointing the same way. You can rotate the blade by reducing its tension and turning these tabs, to change the direction in which the saw cuts.

Truck Design

I want the fabrication process for the monster truck to be as easy as possible, so I'm specifying pine for the body and the wheels. If you try to make it from hardwood, you'll be sawing for a very long time.

The basic plan is shown in Figure 13-5, although naturally you can redraw it to suit yourself. A fanciful 3D rendering is shown in Figure 13-6, and Figure 13-7 is an exploded view, showing how the pieces relate to each other.

When you consider the most visible feature that distinguishes a monster truck from a regular pickup truck, the answer is obvious: the big wheels. So, the first question is how you're going to make those wheels. Maybe there's some really large round dowel that you can cut into slices?

Unfortunately, the largest dowel I've been able to find is 2" diameter, which really isn't big enough.

I think you're going to have to make your own wheels.

Wheel Centering

You need to accomplish two tasks relating to the wheels: draw a circle on your one-by-six board, and mark the center of the circle. You have to know where the center is, so that you can drill it out to fit on an axle.

If you simply draw a pencil line around a circular object, you'll get a circle, but you won't know where the center is. There are some methods to figure this out, but I think the best option is to forget about drawing around an object. Use a piece of cardboard with two holes in it, as shown in Figure 13-8.

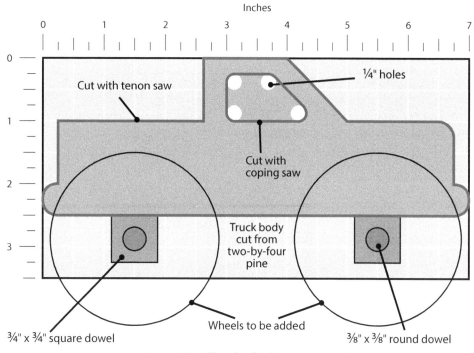

Inches

Cut with tenon saw

¼" holes

Cut with coping saw

Truck body cut from two-by-four pine

Wheels to be added

¾" x ¾" square dowel

⅜" x ⅜" round dowel

Figure 13-5. The plan for the monster truck.

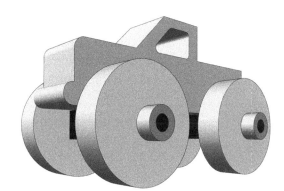

Figure 13-6. Maybe your truck will look something like this.

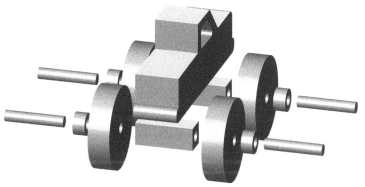

Figure 13-7. Exploded view of the parts of the truck.

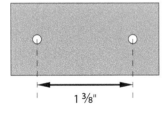

Prick holes
through cardboard
by using an awl.

1 ⅜"

Figure 13-8. Prick two small
holes in a piece of cardboard.
Any cardboard will do.

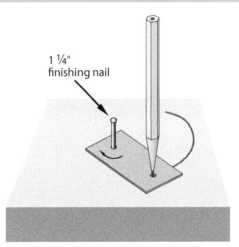

1 ¼"
finishing nail

Figure 13-9. Using the world's simplest
circle-drawing system.

Figure 13-10. After you draw the circle, you can
drill a ⅜" hole in the center.

After you place the cardboard on your one-by-six board, you attach it by using a small nail in one hole. Tap it in no deeper than ½", so that you'll be able to pull it out easily. Now place the tip of your pencil in the other hole and rotate the pencil around the nail, as shown in Figure 13-9.

I'm suggesting that you space the holes in the cardboard 1⅜" apart. This will create wheels that are 2¾" in diameter. If you want wheels that are larger, that's up to you, but you'll have to do more work cutting and sanding them.

After drawing your circle, pull out the nail. You can use the little hole it made as the starting point for drilling a ⅜" hole, because the wheel is eventually going to fit on a ⅜" round dowel as the axle, and drilling the hole is easier before you cut the wheel out of the wood.

You know the procedure to create a hole of this size: clamp the wood to a sacrificial piece, apply a countersink, then use a small bit (maybe ⅛"), then a ¼" bit, and finally the ⅜" bit, running it very slowly and pressing very gently, so that it doesn't run away with you. The result is shown in Figure 13-10.

Wheel Cutting

Now it's time to start cutting around the circle, using your coping saw. This is a very old design of tool that requires good hand-eye coordination and some patience. You may want to test it on some scrap wood, to get the feel of it. I find it easiest to control when I use one hand on the handle, and my other hand stabilizing the frame.

Clamp your board to the work area with about 12" of it sticking out, so that you can cut around the circle without the saw bumping into the clamps. Steer the blade as you follow the circle, as in Figure 13-11. Note that you shouldn't try to turn the blade unless it is moving, so that it has a chance to cut its way into a turn.

While cutting, *pause frequently*. This is hard to do, because I'm sure you want to see the finished result. But the coping saw can easily change its angle without you realizing it. You need to check it every minute or two, and blow sawdust out of the way frequently, so you can see the location of the blade. Always cut a fraction outside of the line, bearing in mind that you can sand it to the precise size afterward.

When you have cut about halfway around the circle, back the blade out of the cut where you started.

Now turn the saw so the teeth face the opposite direction and work it back into the cut with the back side of the blade leading the way. Once you have reached the outline of your circle, begin your second cut as in Figure 13-12. Continue to follow the line until the circle is complete.

Wheel Smoothing

When you finish sawing and extract your wheel, you'll find that it isn't precisely circular, no matter how careful you were. What can you do about this? You can use your electric drill to rotate the wheel against some coarse sandpaper.

To accomplish this, you must mount the wheel on a piece of ⅜" dowel.

Begin by taking a piece of scrap two-by-four and drilling a ⅜" hole all the way through it. This will serve as a very simple jig. Clamp the scrap wood to your work area, then take a ⅜" circular dowel and cut a 2" piece.

Push this piece into the hole in your scrap wood, and grip it with some pliers to keep it under control. Now you can drill a hole in the center of one end with a ⁵⁄₆₄" bit, as in Figure 13-13. The hole should be about ½" deep.

You need to mount the wheel on this dowel in such a way that it fits really

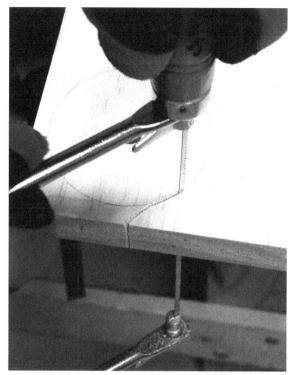

Figure 13-11. Cutting around the circle with a coping saw.

Figure 13-12. Cutting the second half of the circle.

tightly. The way to do this is by driving a screw into the end of the dowel, to make it expand. This is the same principle used in drywall and masonry anchors, which I talked about in the previous project.

The dowel won't expand much unless you cut a slit in the end of it, as in Figure 13-14.

You'll need to wiggle the knife a little to push it into the wood. Keep your other hand well away from the blade. The slit should be about ½" deep.

Pull the dowel out of the scrap wood, and push the split end into the center of your wheel. Insert the other end of the dowel into the chuck of your drill, making sure that the drill is unplugged (if it is a corded drill) or locked, or has its battery disconnected (if it is a cordless drill).

Figure 13-13. Drilling a hole in the center of the ⅜" dowel.

Figure 13-14. Cut a slit in the end of the dowel.

Figure 13-15. Making the dowel expand, so that it fits tightly in the wheel.

Figure 13-16. Run the drill at medium speed while sanding the wheel.

Tighten the chuck. Now you should be able to drive a #8 wood screw into the hole that you drilled in the dowel, as in Figure 13-15. This will force the dowel to expand, so that it is a tight fit in the wheel.

Place a rag on your work area, folded over a couple of times, because you're going to need a shock-absorbing layer for the next procedure. Make sure the rag has no loose ends, so there is no risk of it getting caught in the spinning chuck of your drill.

Clamp some 80-grit sandpaper onto the rag, and you can sand your wheel as in Figure 13-16. This is a much quicker and more accurate way to smooth the wheel than if you tried to manipulate it by hand.

The wheel doesn't have to be perfectly centered around the axle. It just needs to look reasonably round, and you may want to take off the sharp edges by angling the drill.

When you are satisfied with the appearance, loosen the chuck and remove the wheel.

To remove the ⅜" dowel from the center of the wheel, first unscrew the #8 screw to release the pressure in the dowel. Then hold the wheel between your hands with the dowel sticking out downward. Press the bottom end of the dowel against your work area, push down on the wheel, and the dowel will slide up and out.

After you've done all that—you just need to repeat the whole procedure, to make three more wheels!

Cutting the Truck Body

The body of the truck requires a piece of two-by-four that is 7" long, without any major defects in it.

Before you cut around the outline of the truck, you have to decide if you want windows. Making side windows will be relatively easy: just drill through the wood with a ¼" bit as in Figure 13-17, and then use your coping saw to modify the opening, as in Figure 13-18.

The finished window is shown in Figure 13-19. To smooth the edges, you can wrap sandpaper around

Figure 13-17. These ¼" holes were located by pricking through the plan for the truck.

Figure 13-18. Using the coping saw to connect the holes.

Figure 13-19. The window, with rough edges.

Figure 13-20. Clamp the two-by-four so that you can cut down into the wood.

Figure 13-21. Making the second cut. The wood you are cutting will vibrate less if the saw is working as near as possible to the horizontal clamp. You will have to unclamp the wood and move it up a couple of times, as you progress.

Figure 13-22. Ready to cut the slope of the windshield.

a thin object such as a pencil, and poke it through the hole. Personally, I prefer to use metal files for this kind of task. A semicircular file and circular file work well for me. To smooth straight edges of softwood, I use a flat file. The correct tool for this job is a **rasp**, but most rasps are relatively expensive and tend to be too coarse for this kind of detail finishing of softwood.

Now you can use your tenon saw to make straight cuts in the two-by-four. First cut along the bottom of the body of the truck. After you have done that, I suggest you should turn the wood so that you are cutting downward, making it easier to check that the saw is horizontal and the cut that it makes is vertical. The first cut of this kind is shown in Figure 13-20.

The remaining cuts will be easiest if they, too, can be made by sawing down vertically into the wood. One way to hold the two-by-four is shown in Figure 13-21, where a cut has just been started. A piece of scrap wood is clamped to the work area, and the two-by-four is clamped to the scrap. If you tighten the clamps as much as you can, I think you will find that the wood doesn't slip.

The windshield area of the truck is angled at 45 degrees. This cut, also, will be easiest if you turn it so that it is vertical. See Figure 13-22.

After you have cut all around the body of the truck, you can consider an interesting question. Do you think it would look better if you drilled another hole to make the windshield, extending all the way through to make a rear window, too?

This would be challenging, because you're working with soft pine, which splits easily. The more you remove from it, the weaker it gets. Personally, I decided to skip the windshield. You can give it a try if you like, but be sure to have a spare piece of two-by-four ready if your work piece splits unexpectedly.

Truck Assembly

It would be great if we could give your truck coil-spring suspension, but I'm afraid that isn't an option. The suspension will consist of ¾" square dowel, into which you will glue axles consisting of ⅜" round dowel.

Figures 13-23 through 13-26 show the steps to assemble a pair of wheels. In Figure 13-23, a piece of square dowel has been drilled with a ⅜" hole at each end. Pieces of ⅜" dowel will be glued into the holes.

In Figure 13-24, the axles have been glued in, and the wheels are ready to be slipped onto the axles. The wheels will be able to rotate freely.

In Figure 13-25, the wheels are on the axles, but we need to stop them from coming off. For this purpose, a retaining ring will be glued onto the end of each axle.

Finally the wheel-and-axle assembly is ready to be mounted under rear of the truck, as shown in Figure 13-26. An identical assembly will be mounted under the front of the truck.

Now that you see the plan, you should have no trouble making it. First cut two 2" pieces of the square ¾" x ¾" dowel, clamp one of them vertically, and mark its center as in Figure 13-27.

Now you need a ⅜" diameter hole, ¾" deep. Naturally you will need to drill

Figure 13-23. Two ⅜" axles will be glued into holes at the ends of a ¾" x ¾" square dowel.

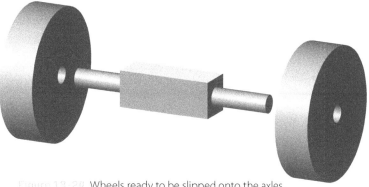

Figure 13-24. Wheels ready to be slipped onto the axles.

Figure 13-25. Retaining rings ready to be glued in place.

Figure 13-26. Rear-wheel assembly has been mounted under the truck.

Figure 13-27. Square dowel, clamped to a block of scrap wood.

Figure 13-28. Drilling a ⅜" hole, ½" deep, in the end of the square dowel.

smaller holes before using the ⅜" bit, and you'll need to mark the depth on the drill bits with some masking tape. Assuming you have worked your way through previous projects, you already know how to do this. See Figure 13-28.

Drill another hole at the opposite end of the square dowel, and then make similar holes in the second 2" piece of square dowel.

For the axles, you need four pieces of round ⅜" dowel, each 2" long. Glue them into the holes that you drilled in the ends of the dowels.

When the glue has set, it will be easiest to attach the square dowels to the underside of the truck before you add the wheels. You can use 1½" #8 wood screws to do this.

First drill two screw holes in the side of a dowel, centered, ¾" apart, using a ⁵⁄₃₂" bit. Countersink each hole, then clamp each square dowel to the underside of the truck and use a ⁷⁄₆₄" bit, through the screw holes, to make pilot holes in the truck body.

Attach the square ¾" dowels using the wood screws, and then slide the wheels onto the axles. Now you need a way to make the retaining rings for the wheels, which can be done by taking a piece of ¾" round dowel, 2" long, and drilling a ⅜" diameter hole all the way through the center—exactly as you did when making the Swanee whistle in Chapter 9. Saw the round ¾" dowel into four slices at ½" intervals. Each slice can be glued onto the end of an axle—after you put the wheel on the axle, of course. I suggest using epoxy glue, because the retaining rings may not make a very tight fit, and cannot be clamped. Be careful not to get any glue between the wheels and the axle, which would stop them from turning.

The truck that I made is in Figure 13-29. It doesn't have a high quality of finish, because basic hand tools don't really enable this when shaping soft pine.

My only concern about this project is that it uses 1½" screws to attach the square ¾" x ¾" dowels under the truck. Conceivably, an infant could throw the truck hard enough to split it open, in which case, could the screws be hazardous?

If you are concerned about this, you can easily modify the design. Instead of 1½" screws, use ¼" pegs as you did in the ends of the shelves in the bookcase, in Chapter 11. This will take a bit longer, which is why I didn't specify it as the default procedure in the project.

Figure 13-29.
The finished truck.

Other Ideas

The ability to saw along curves suggests many other projects.

Wine Rack

How about the kind of wine rack in which the bottles are inserted neck-down? All it takes is a length of hardwood board and some holes big enough for the necks of the bottles. If you want to hang it on the wall, you'll need a couple of brackets to support each end. See Chapter 12 for details about attaching a shelf to a wall.

Four Across

Fabricating your own version of the four-across game would be challenging, and would require a Forstner bit to drill smooth, regular holes. See the Holes and Curves Fact Sheet on page 164 to learn about Forstner bits.

Four Across is a game where players take turns dropping white and black discs down vertical channels, trying to be the first to get four of one color in a horizontal row. You can find pictures of it easily if you search online.

You would have to mount vertical internal strips to separate the channels, but that could be done with ⅜" square dowel, which is available from crafts stores. For the counters, you could either cut slices of 1¼" round dowel or look online, on a source such as eBay, for plastic counters or discs. Many sizes are available. You should obtain the counters before you build the game, to make sure they will fit.

This would not be a trivial project, but the result could look really nice.

Holes and Curves Fact Sheet

There are two different types of holes that you may want to cut in a piece of wood: round holes, and holes that are not round. I'll deal with the round holes first.

Large, Clean Round Holes

While a handheld drill can be used with most of these hole-cutting devices, a drill press is really the right tool for the job. See page 243 in Chapter 20.

Hole Saws

A hole saw is like a saw blade curved to form a continuous circle. A typical kit of hole saws is shown in Figure 13-30. Each saw screws onto a hub that has a six-sided shaft. The shaft engages in the chuck of a drill.

Using a hole saw larger than ¾" in a hand-held drill can be challenging, because of the backlash when the saw digs in. Make sure the wood is securely clamped, and hold the drill firmly with both hands. Hole saws

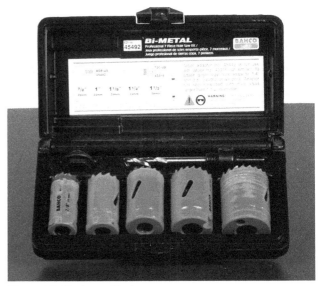

Figure 13-30. A set of hole saws.

larger than 1½" may require more power than a cordless drill can provide. The type of drill that has a side handle is preferred. Use a gentle touch, and don't run the drill too fast.

After you cut each hole, the hole saw ends up with a disc of wood stuck in it. Slots in the side of the hole saw enable you to stick the end of a small screwdriver in, to push the disc out, but this is an annoying chore. Another way to deal with it is to turn the hole saw so that its teeth face upward, clamp it on your work area, and drive a 3" screw down into.the disc of wood. When the screw hits bottom and keeps turning, its thread will raise the disc up and out.

Hole saws are available with a variety of cup depths.

Forstner Bits

For sizes up to 2" diameter, I think **Forstner** bits are unbeatable. Each bit is like a miniature hole saw with two radial blades. The blades remove the wood that would normally clog a hole saw. A set of Forstner bits is shown in Figure 13-31.

Forstners create very clean holes in sizes from ¼" upward, in increments of ⅛". The larger sizes are more difficult to control, and may overload a cordless drill.

Because a Forstner bit removes debris as it descends into the wood, you can stop at any point, leaving yourself with a hole that has a flat bottom with a dimple in the middle. You can also do this with a spade bit (described below), but the hole may not be as neat and clean.

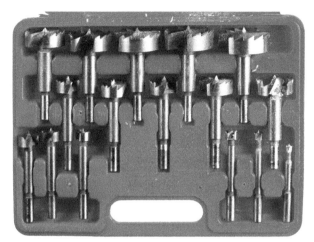

Figure 13-31   A well-used set of Forstner bits.

The disadvantage of Forstner bits is that they are more expensive than other methods for making circular holes.

Spade Bits

These are often used by electricians to make holes in two-by-fours when running cables through a wall. You won't get a neat edge, but spade bits are affordable and quick to use. Controlling them requires some strength and practice. Figure 13-32 shows some spade bits.

Safety note: hole saws, Forstner bits, and spade bits are all potentially dangerous, because the cutting edges are exposed without any kind of guard. They are especially hazardous in a handheld drill. If the bit jams in the wood, the drill can wrench itself free from your hands, with unpredictable consequences.

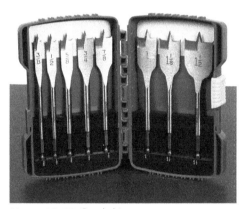

Figure 13-32   Spade bits.

Adjustable Hole Cutter

For holes larger than 2½", I think this is your best option, shown in Figure 13-33. It consists of a central ¼" drill bit that centers and stabilizes the cutter, and a horizontal bar with two blades that can be moved inward and outward. The blade positions must be set manually, but can be adjusted for precisely the size of hole that is desired.

A variant from Milwaukee has preset blade positions, the advantage being that you don't need a hex wrench to loosen and tighten the blades. It is marketed specifically for cutting holes in a ceiling where reflector lights will be installed.

Don't even think about using this device to cut wood with a handheld drill. The speed of each blade relative to the material increases radically when the blades are further out from the central bit. An adjustable hole cutter should only be used in a drill press at its lowest speed, with the material secured by at least two clamps. Even then, there is the possibility of it splitting the wood and throwing pieces in unpredictable directions.

Figure 13-33 An adjustable hole cutter.

Holes That Are Not Round

The time-honored method for cutting an irregular-shaped hole is to drill a few holes with a normal bit, then use a saw with a blade that is narrow enough fit into one of the holes and go around corners. Saws of this type include **coping saws, scroll jaws, keyhole saws, and jigsaws**.

Coping Saw

This is the saw recommended for the Monster Truck project. Instructions for using it will be found beginning on page 153.

Scroll Saw

The hand-operated version of this tool looks like a coping saw, but with a deeper frame, allowing you to cut in further from the edge of a piece of wood.

The powered version has an oscillating blade above a flat table. You cut into a piece of wood from the edge, or remove the blade and thread it through a hole in exactly the same way as the blade in a hand-operated scroll saw or coping saw. The great advantage of the powered scroll saw is that it leaves both hands free to push or rotate the material that is being cut. However, the exposed blade is a safety hazard. If you make a hasty or uncontrolled hand movement, an injury can result.

Keyhole Saw

This is a hand saw with a small, tapering blade that is just narrow enough to fit into a ⅜" hole. Because the blade is open at one end, you can use it at any distance from the edge of the wood. The tradeoff is that accurate cuts are difficult to make. A very low-cost keyhole saw is shown in Figure 13-34.

Figure 13-34: A keyhole saw with a replaceable blade.

Jigsaw

An electrically powered jigsaw is a handheld device with a blade that sticks out at the bottom, as shown in Figure 13-35. Insert the blade in a hole, and the saw cuts in whichever direction you are pointing it. Blades are interchangeable, in various lengths and with various tooth spacing. Longer blades may be wider to withstand the force when you are cutting materials of ¾" or thicker.

Because the blade is only supported at the top end, the lower end can flex from side to side, especially if you are cutting a curve that turns across the grain. The result will be a hole that does not have vertical edges. Care and skill are needed to control a jigsaw accurately.

Figure 13-35: A battery-powered jigsaw.

Battery-powered jigsaws tend to be less powerful than plug-in models.

The exposed blade of a jigsaw makes it yet another safety hazard. Never put your fingers near the blade, and leave the switch of a jigsaw in its locked position, especially on a battery-powered model.

All cutting tools have unique risks associated with them. Read the user manuals carefully, and seek some in-person guidance from a skilled person if you are young or inexperienced.

- All about nuts, bolts, and washers
- Choosing between bolts and wood screws
- Linkages, pivots, and friction

YOU WILL NEED

- Two-by-four pine, no knots, not warped, length 12"
- Square dowel, hardwood preferred, ¾" x ¾", length 12"
- "Hobby board," maple, ¼" thick, 3½" wide, length 36"
- Wood screws, #6 x ⅝", flat head, Phillips, quantity 4
- Wood screws, #10 x 1¼", round head, straight slot, quantity 6
- Bolts, ¼" x 1", 20 tpi, hex or Phillips pan head, quantity 2
- Bolts, ¼" x 1½", 20 tpi, partially threaded, hex head, quantity 2 (optional)
- Locknuts, nylon inserts, ¼", 20 tpi, quantity 2
- Fender washers, ¼" center hole, 1" diameter, quantity 18

Continued on the following page

Do you want a device that makes a copy of a simple drawing, and enlarges it at the same time, while using no electricity at all? A gadget that consists of just a few pieces of wood, some screws and bolts, a pointer, and a pen?

That device is a *pantograph*, which will demonstrate some important concepts: rotating **linkages**, and **nuts** and **bolts** that connect them.

Nuts and Bolts for This Project

Check the Nuts and Bolts Fact Sheet on page 180 if you need basic information.

For this project, you can use ¼" size bolts that are 1" long with 20 threads per inch. You may see them listed in catalogs as ¼-20. The nuts must also be ¼-20, to match the bolts. You only need two nuts and two bolts, and can probably find small quantities in little plastic bags on hooks in your local hardware store.

You will also need **washers**. A basic washer is just a stamped circle of metal with a hole in the middle. Its purpose is to spread the force exerted by a nut or the head of a bolt, and also to reduce wear if the nut or the bolt rotates against a soft surface—such as a piece of wood.

I'd like you to buy **fender washers** such as the one shown in Figure 14-1. This type of washer is wider than a standard washer. The ones I am suggesting are described as ¼" size (to fit the bolt) and 1" in diameter. If you can only find ¾" diameter, they'll do.

Figure 14-1. Fender washer, 1" diameter, for ¼" bolts.

If you are willing to take a little extra trouble, some refinements are possible.

Instead of generic nuts, you can buy **locknuts** with **nylon inserts**. The insert grips the thread of the bolt, and stops the nut from working loose.

Instead of generic 1" bolts, you can buy 1½" **partially threaded** bolts. These

Figure 14-2. A partially threaded bolt (length 1½", size ¼").

resemble wood screws because they have a smooth shank below the head, and the thread doesn't start until halfway down. An example is shown in Figure 14-2. The ones you should buy must have 20 threads per inch, to match the nuts.

I prefer this type of bolt for this project because each bolt will be inserted through two pieces of wood that rotate around it. If the bolt is threaded all the way up, the thread will tend to eat away at the wood, and the joint will become loose. A smooth shank at the top of the bolt inflicts much less wear—just like the smooth shank of a wood screw. The principle is illustrated in Figure 14-3.

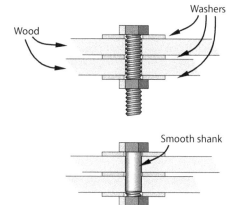

Partially threaded bolts are also known as **partially threaded hex screws**, when they have hexagon-shaped heads. For this project, ideally they should be ¼" diameter and 1"

Figure 14-3 When a softer material such as wood rotates around a bolt, a smooth shank reduces wear.

long, with half of the body threaded and the other half smooth. Unfortunately I have never seen this specification, so I used bolts 1½" long with about ¾" of the body threaded, as in Figure 14-2. They are longer than necessary, but they'll do.

If you don't want the hassle of finding this type of bolt, and you don't want to pay for a substantial minimum quantity, use generic 1" bolts that are conventionally threaded.

Head Profiles

Larger bolts mostly have hexagonal heads. Smaller bolts may have other styles of heads, similar to the range available for wood screws and sheet-metal screws.

A selection of screws is shown in Figure 10-33 on page 123, while a selection of bolts appears here in Figure 14-4. They range from one 1" long with a ½" diameter

Finishing nails, 1¼"
Wrench, adjustable, length 4" or 6"
Screwdriver, medium size, flat blade

Also, as listed previously: Tenon saw, miter box, trigger clamps, ruler, speed square, rubber sanding block, work gloves, awl, masking tape, pliers, electric drill and drill bits, countersink, screwdriver or electric screwdriver, dust mask (optional), safety glasses (optional), utility saw (optional), plywood work surface, sandpaper, and epoxy glue and hardener.

Check the Buying Guide on page 248 for information about buying these items.

A variety of bolts.

thread, at far left, to a #3 size bolt ¼" long, at far right. From left to right, the first two have hex heads, followed by flat head with straight slot, round head with straight slot, pan head with Phillips slot, flat head with Phillips slot, and pan head with Phillips slot. The plated steel bolts have a slightly blue tint, while the two pan-headed bolts look slightly brown, which is characteristic of stainless steel.

For this project, the two bolts and the screws that are #10 size, 1¼" long must not have flat heads, because I don't want them to be recessed into the wood. A countersink removes some wood, and I want as much wood as possible to be in contact with the shanks of the fasteners.

For the wood screws, you will be looking for either pan heads or round heads. Round heads seem to be more common than pan heads in wood screws, and often have straight screwdriver slots instead of a Phillips head. Any type with an underside of the head that is not beveled will be okay.

Tools for Nuts and Bolts

I'm going to assume that you may not have any wrenches for working with nuts and bolts, so here's a quick primer.

A nut has an internal thread that matches the thread on the bolt, so that if you hold the bolt stationary while you turn the nut, the nut moves up the thread toward the head of the bolt, and can exert a powerful gripping force on anything in between.

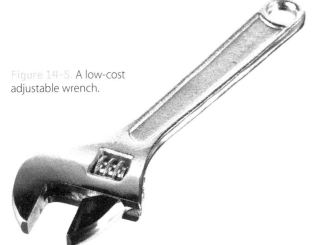

Figure 14-5. A low-cost adjustable wrench.

To make this happen, you need a tool that is stronger than your fingers. A **wrench** is the obvious choice, and a small **adjustable wrench** is the cheapest way to go. It does have disadvantages: it only grips two of the six sides of a nut, and tends to slip off, but it should be good enough for the projects in this book, and you can buy one online for less than the price of a hamburger (without fries). An example is shown in Figure 14-5. Other alternatives for gripping and turning hexagon-shaped nuts and bolt heads are described in the Nuts and Bolts Fact Sheet on page 180.

If you want to turn a nut while preventing the bolt from turning (or vice-versa), you need a way to grip both of them at the same time. This is not a problem if your bolt has a Phillips head: you immobilize it with a Phillips screwdriver, while you use your wrench on the nut. However, if you have a bolt with a hex head, you now have two hexagon-shaped things to grip, and you only have one wrench.

The easy answer is to use pliers on the part that you want to prevent from turning. This is not the right thing to do, because pliers tend to slip, and when they slip, they chew up the metal. But if you don't expect to do a lot of work with nuts and bolts in the future, pliers will be okay for the projects in this book.

Choosing the Wood

Your pantograph will be fabricated from four strips of wood, each of which is 16½" long, ¼" thick, and about 1½" wide. You could use plywood, but for this project, I prefer something with harder edges.

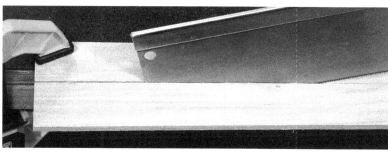

Figure 14-6: When slicing a strip of poplar parallel with the grain, no sacrificial wood is needed.

At my local big-box store I found so-called "hobby board" made of ¼" poplar. This is not the hardest wood, but hard enough. The strips are 3½" wide, but slicing them down the middle to make narrower strips does not require much work, because the wood is so thin. Also, when you're cutting parallel with the grain, the saw doesn't create significant splinters, and you don't need sacrificial wood. See Figure 14-6.

You can search online for narrower strips that don't require any cutting, but for me, the hobby board was a quick and simple solution. Here's the procedure:

- Buy at least 36" of ¼" x 3½".
- Cut two lengths out of it, each measuring 16½" long and 3½" wide.
- Slice each section into two pieces, each measuring 16½" long and slightly more than 1½" wide. (Making a long cut like this, with the grain, is known as ripping the wood.)
- You end up with four strips. Use your pencil to number the strips 1, 2, 3, and 4.

The length of each strip is important, but the exact width is not.

Identifying the Parts

Figure 14-7 gives you an idea of what a pantograph looks like. In this diagram, someone is using it to trace a simple drawing and make a copy that is 1.5 times as big. Note that the arms of the pantograph have holes in them (which penetrate all the way through). The holes allow the arms to pivot around screws and bolts. Extra holes allow configurations that will magnify the original drawing by a factor of 1.5, 2, 3, or 5.

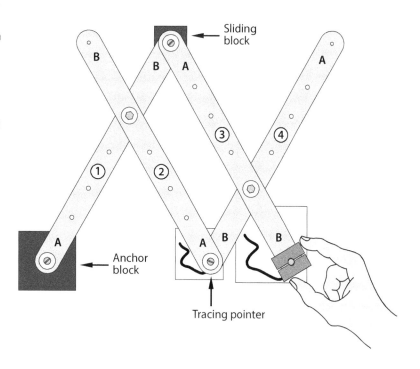

Figure 14-7: A pantograph being used to make a copy 1.5 times as big as the original.

Sliding block

Anchor block

Tracing pointer

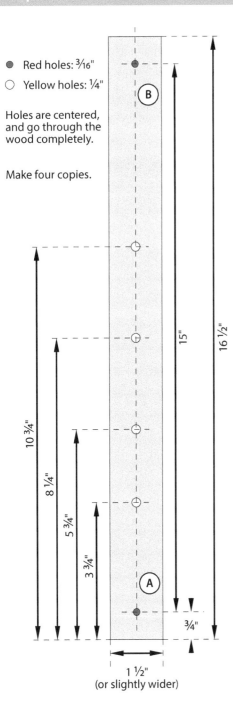

Red holes: 3/16"
Yellow holes: 1/4"

Holes are centered, and go through the wood completely.

Make four copies.

B

15"

16 1/2"

10 3/4"

8 1/4"

5 3/4"

3 3/4"

A

3/4"

1 1/2"
(or slightly wider)

Figure 14-8. Locations of holes to be drilled in each of four pieces. Don't forget to label the ends A and B.

Drilling the Holes

Figure 14-8 shows the sizes and locations of the holes that you need to drill. The locations are identical in all four of the pieces. Even though you're using hardwood, you should take some trouble to drill into sacrificial wood below it, as I have described in previous projects. Rough-edged, splintery holes will interfere with the assembly of this project.

I suggest using an awl to mark each location, then drill all of the locations with a 1/8" bit, then all of them with a 3/16" bit. Finally enlarge the four holes in the middle to 1/4", while the holes at the ends remain 3/16".

Using a countersink is not a good idea, because I want to maintain the full thickness of the wood around each hole.

Perhaps you are thinking that since the holes are in the same positions in each strip, you could stack the strips, clamp them, and drill through all of them at once. Yes, you could do that, but if your drill isn't precisely vertical, the spacing of the holes in the lower pieces of wood will be inaccurate. I think it's best to drill one strip at a time.

After you drill the holes in each strip, label the ends A and B as in the diagram.

Each end labeled A has a nearest hole 3" away. Each end labeled B has a nearest hole 5" away.

Throughout the text, I'm going to refer to the strips of wood as Arm 1, Arm 2, Arm 3, and Arm 4, as shown in Figure 14-7. Refer back to that figure while you are building this project, because identifying the arms and getting them the right way around is essential. Overlapping them correctly will also be important.

In Figure 14-7, the arms are shown with rounded ends. You'll find in the photographs throughout this project that I rounded the ends of the arms in my pantograph. Whether you bother to do this is up to you. It makes no functional difference.

Fabricating a Pencil Gripper

The output from the pantograph is created with a pen or pencil that has to be gripped by some kind of gadget at the end of Arm 3. I'll deal with this first.

I suggest you begin by paging ahead to Figure 14-23, so that you can see what the pencil gripper looks like when it is complete. Then the steps along the way will be a little less mysterious.

Select Arm 3. You're going to operate on the B end, so make sure it's the right way around. Figure 14-9 shows the plan for drilling this arm. The gray lines are pencil lines.

Figure 14-10 shows the actual work piece. First draw lines across the arm using a speed square, then use your awl to make two prick marks, each ⅜" from the center line.

The next step is to drill a ⅛" screw hole on each prick mark, all the way through, and use a countersink to bevel these holes, because they are not going to rotate around bolts or screws. See Figure 14-11.

Enlarge the end hole in Arm 3 from ³⁄₁₆" to ⅜" diameter. If you are going to be using a pen that is fatter than ⅜", you need to enlarge this hole even more. Since ⅜" is your largest drill bit, you can use a ½" countersink (pushing it all the way through the wood) if necessary. See Figure 14-12.

The last step in this sequence is to saw along the line that crosses the hole, as in Figure 14-13.

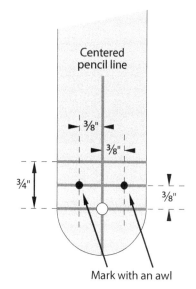

Figure 14-9. The plan for the B end of Arm 3.

Figure 14-10. Arm 3 with awl marks ⅜" either side of the center line.

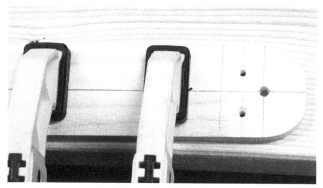

Figure 14-11. Arm 3 with beveled ⅛" holes.

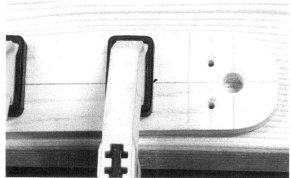

Figure 14-12. The hole at the end of the arm has been enlarged to ⅜".

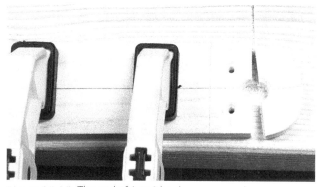

Figure 14-13. The end of Arm 3 has been removed.

Pencil lines on
¾" x ¾" square dowel

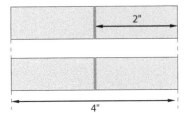

Figure 14-14. Mark the midpoints of two dowels.

Now set aside Arm 3 and cut two pieces of ¾" x ¾" square dowel, each 4" long. Use your speed square to draw a line marking the midpoint of each one, as shown in Figure 14-14.

Clamp the two dowels and drill a ⁵⁄₃₂" hole where they touch each other, as in Figure 14-15. If they are clamped firmly, the drill shouldn't push them apart. Enlarge this hole to ⁷⁄₃₂" and finally ¼". This is going to be your pencil gripper.

After you release the dowels, they should look like Figure 14-16.

Clamp one of the dowels under the end of Arm 3 as in Figure 14-17, so that the edges are aligned, the centers are aligned, and the semicircular notches are aligned. In the screw hole that is exposed, drill a ³⁄₃₂" pilot hole down about ½" deep in the dowel.

Insert a #6 size, ⅝" screw as in Figure 14-18.

Figure 14-15. Making a hole where the dowels touch each other.

Figure 14-16. This will become your pencil gripper in the pantograph.

Figure 14-17. One of the dowels is clamped under the end of Arm 3. Drill a ³⁄₃₂" pilot hole through the screw hole, ½" into the dowel.

Figure 14-18. A screw has been inserted in the pilot hole.

Now swap the clamps around to expose the other hole, drill a pilot hole there, and insert another #6 size, ⅝" screw.

When you've done this, go back to your other piece of dowel. In Figure 14-19, notice that the notch that you drilled in it is facing downward. Make two marks with your awl, each ⅝" from the center line. Then drill screw holes at these locations, all the way through the dowel, with a ³⁄₁₆" bit.

It's time to assemble your pencil gripper. Place the first dowel horizontally with Arm C hanging down from it, and put the second dowel on top of it, as in Figure 14-20. Clamp them in place and use a ⅛" drill bit through the screw holes to make large pilot holes in the lower dowel, about ⅝" deep.

Insert two #10 screws, 1¼" long, as in Figure 14-21. Maybe you can see how this is going to work, now: If you loosen the screws, you can insert a pencil in the hole between the dowels, and when you tighten the screws, they'll grip the pencil. For the time being, don't insert a pencil, just tighten the screws.

You can trim 1" off the ends of the dowels, because they only need to be 2" long. I asked you to make them 4" long because you needed the extra wood for clamping in the previous steps. Figure 14-22 shows the dowels being trimmed.

Finally, Figure 14-23 shows the completed pencil gripper at the end of Arm 3. You've finished working on Arm 3 for the time being, so you can set it aside.

Figure 14-19. In the second dowel, each awl mark is ⅝" from the center line. The notch in the dowel points downward. Drill holes with a ³⁄₁₆" bit.

Figure 14-20. Use a ⅛" drill through the screw holes to make pilot holes in the lower piece of dowel.

Figure 14-21. Insert two 1¼"long, #10 screws.

Figure 14-22. Trimming 1" off the dowels.

Figure 14-23. The completed pencil gripper.

Figure 14-24. Location of a ⅛" beveled screw hole in Arm 4 at the B end.

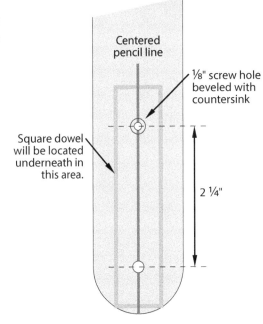

Centered pencil line

⅛" screw hole beveled with countersink

Square dowel will be located underneath in this area.

2 ¼"

Figure 14-25. A 3½" piece of dowel clamped under Arm 4.

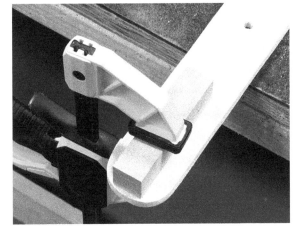

The Pointer Block

The pantograph uses a pointer to trace original art. It will be a 1¼" finishing nail glued into a piece of ¾" x ¾" square dowel, and the dowel will be mounted underneath the end of Arm 4. You're going to operate on the end of Arm 4 identified as B. Make sure you're not dealing with the other end by mistake.

Drill a ⅛" screw hole positioned 2¼" up the center line from the hole at the bottom of Arm 4, as shown in the diagram in Figure 14-24. Bevel the hole, as you did with the holes that you drilled in Arm 3.

Turn Arm 4 over, and clamp a piece of ¾" x ¾" square dowel to it, as in Figure 12-25. The dowel is 3½" long, almost reaches the end of the arm, and is centered between the two edges.

Flip the arm over again (keeping the clamp attached), and drill a ³⁄₃₂" pilot hole by inserting the bit through the screw hole that you drilled previously. See Figure 14-26. Then drive a #6 size, ⅝" screw through the screw hole into the dowel underneath, as in Figure 14-27.

Move along to the end of Arm 4, and use a ⅛" bit to drill through the ³⁄₁₆" hole, about ½" into the dowel beneath. This ⅛" hole is a pilot hole for a 1¼" long, #10 screw, which you are going to insert next. See Figure 14-28.

Take Arm 2 and select end A. In Figure 14-29, the round-headed screw is being inserted through a washer, through Arm 2, through another washer, through Arm 4, and into the pilot hole that you just made in the square dowel. Arm 2 will pivot around this screw. Check back to Figure 14-7 for an overview. Tighten the screw, but then slacken it by a quarter-turn, so that the arm pivots easily. The pantograph still needs a pointer to be added to the square dowel, but not just yet.

Set aside Arm 2 and Arm 4. It's time to deal with Arm 1. End A of it is going to be attached to a block of two-by-four which you should cut now, measuring 3½" x 3½" square. Draw an X connecting its corners, to find the center. Drill a ⅛" guide hole in the center that is ¾" deep, and you're ready to join Arm 1 to it, as shown in Figure 14-30. A 1¼" long, round-headed screw will go through a washer, then through the A end of Arm 1, and through another washer, into the guide hole in the two-by-four. Once again, tighten the screw but then slacken it off by a quarter-turn.

Cut another block of two-by-four, this time 2" x 2" square. Mark the center, drill a ⅛" guide hole ½" deep, and attach the B end of Arm 1 and the A end of Arm 3, as shown in Figure 14-31, with Arm 3 overlapping Arm 1. This time, because you have two arms, you need three washers, one under the screw head, one between the arms, and

Figure 14-26. Drill a 3/32" pilot hole.

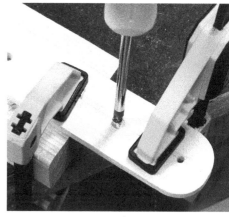

Figure 14-27. Insert a ⅝" #6 screw to hold the dowel underneath.

Figure 14-28. A ⅛" pilot hole at the end of Arm 4.

Figure 14-29. The screw joins Arm 2 with Arm 4.

Figure 14-30. Arm 1 will be anchored to the square block.

Figure 14-31. Assembling arms 1 and 3.

Figure 14-32. Arm 1 and Arm 3 connected.

Figure 14-33. Tightening a bolt to connect Arm 2 and 1.

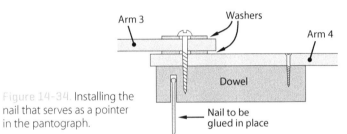

Figure 14-34. Installing the nail that serves as a pointer in the pantograph.

Figure 14-35. Aligning the pencil, the nail, and the anchor block.

one between Arm 3 and the block. The assembly is shown in Figure 14-32. Tighten the screw and then slacken it off by a quarter-turn.

You're almost done. You now have two pairs of arms, and you need to link them using nuts and bolts. Check back to Figure 14-7 to refresh your memory. The bolts in that figure are shown with hex heads.

Lay Arm 2 over Arm 1 and push a ¼" bolt through the holes, with a washer under the head of the bolt, a washer between the arms, and a washer under Arm 1. The bolt may be a tight fit in the ¼" holes, but this is good: you don't want it to be loose. If you can't push it through with your fingers, tap it with a hammer, or screw it through with a wrench.

Attach a nut on the underside, grip it with your pliers, and tighten the bolt with your wrench, as in Figure 14-33.

Arms 3 and 4 are connected in exactly the same way. Once again, refer back to Figure 14-7 to make sure you have them connected right.

Now, there's just one more thing to take care of, which is to drill a hole for the finishing nail that will function as a pointer in the dowel underneath Arm 4. A ⅛" hole should work. Be careful, though, that you don't drill into the screw that entered the dowel from the other side. This will break your drill bit. Figure 14-34 shows a cross-section side view. Locate the nail hole about ¼" from the end of the dowel. Drill gently, to avoid splitting it.

Mix a little epoxy glue, turn the pantograph upside-down, and dribble the glue into the nail hole. Insert the head of the nail into the glue, and turn the pantograph on its side, as in Figure 14-35. The underside of the large wooden block, the point of the pencil, and the point of the nail should all line up, so that when the pantograph is placed on a table, they will all sit flat and make contact with the table. I used a speed square, as in this figure, to check the adjustment. Then I waited for the glue to set.

Using the Pantograph

Figure 14-36 shows the pantograph in action, replicating a drawing at 1.5 times its original size. You'll notice that the anchor block has been taped to the work surface, along with the original drawing and the paper for the copy.

Using the pantograph is a little counter-intuitive. You don't take hold of the pointer. You *hold the pencil, and watch the pointer* as you move the pencil. You will find, magically, that your pencil is duplicating the art. But why does it work?

How It Works

Check the simplified diagram in Figure 14-37. Arms 1 and 4 of the pantograph must be parallel with each other, and arms 2 and 3 must be parallel with each other. You take care of this by choosing the holes that you use for your two bolts.

It's fairly easy to prove, geometrically, that the ratio between X1 and Y1 will always be the same as the ratio between X2 and Y2. So, your choice of a hole in Arm 1 determines your magnification ratio. The way I set things up, the bolt is in the hole located 10" from the pivot in the anchor block. So, Y1 = 10. We know that X1 is fixed at 15", so the magnification ratio will be 15 divided by 10 = 1.5.

If you unscrew the bolts and move arms C and D to the left, as in Figure 14-38, now you're using the 7½" holes, so the magnification is 2:1. If you move C and D another step to the left, you get 3:1, and finally 5:1.

Breaking the Rules

I said that Arm 1 must always be parallel with Arm 4, and Arm 2 must always be parallel with Arm 3. But—what if you break the rules? Suppose you unscrew the nuts and bolts and connect the bolts through randomly selected holes. What do you think will happen?

You'll find that making the distances unequal will result in an interestingly and strangely distorted copy of the original drawing. Adding distortion is something unique to the pantograph. (A photocopy machine can't do that.)

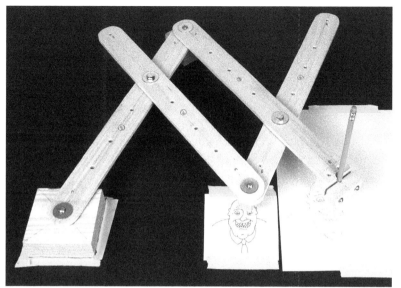

Figure 14-36. The pantograph replicating and enlarging a drawing of a handsome guy.

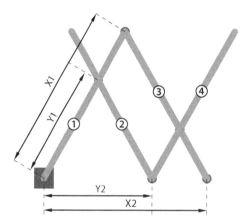

Figure 14-37. How a pantograph magnifies a drawing.

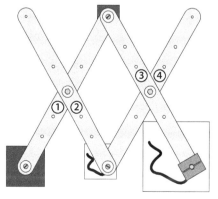

Figure 14-38. Setting the pantograph for 2:1 magnification.

Nuts and Bolts Fact Sheet

The Basics

The majority of nuts and bolts are made of steel, plated to resist rust. Stainless-steel versions are also available, at a slightly higher price, and many special alloys are used in fields such as aerospace where temperatures and stress would tend to damage regular steel fasteners.

Nut sizes are measured in inches (in the United States), from one flat side of the nut to the flat side opposite. The most common sizes are:

- $\frac{3}{16}$"
- $\frac{1}{4}$"
- $\frac{5}{16}$"
- $\frac{3}{8}$"
- $\frac{7}{16}$"
- $\frac{1}{2}$"

The thickness of the body of a bolt is measured with a gauge number. (These are comparable to the gauge numbers for screws. See page 124.) Thicker bolts have their diameters expressed in fractions of an inch.

Larger bolts tend to have fewer threads per inch (abbreviated tpi, not to be confused with teeth per inch in saws). Often you have a choice of tpi for a particular gauge of bolt. For example, a #10 bolt may have either 24 or 32 threads per inch. It will be identified as 10-24 or 10-32. The nuts that you use *must match*, so shop with care!

While the US generally does not use the metric system to measure screws, metric nuts and bolts are relatively common, because the metric system is used in many imported cars and appliances.

Bolts vs. Wood Screws

Bolts are more often used in mechanical equipment than with wood. But not always! How do you decide whether to use a wood screw or a bolt?

If you are attaching something to a piece of wood that is $\frac{1}{2}$" thick or more, a wood screw can usually do the job, especially if the two pieces will remain tightly joined together.

If you are attaching something to a piece of wood less than $\frac{1}{2}$" thick, the thread of a wood screw will not have much to hold onto. A better option is to put a bolt through the pair, and attach a nut to clamp them together.

The pantograph has two pivoting joints consisting of layers of $\frac{1}{4}$" wood, so bolts are the way to go in those locations. But it also contains pivots embedded in wooden blocks, where screws are easy to use.

Retaining a Nut

The nut that you tighten on a bolt can easily become loose in response to flexing, temperature changes, or vibration. Several options can prevent this.

The old-school way is to add a second nut, and tighten it against the first nut. This is often called a **jam nut**. To do this, you need a way to grip the first nut while you rotate the second nut. This is not going to be easy, because the tool holding the lower nut must be as thin as the nut, or thinner. In any case, sometimes the two nuts end up loosening anyway.

Nylon-insert locknuts are a much better way to go. They cost more, but you don't have to worry about them unscrewing themselves.

Many other types of locknuts exist for particular applications. For everyday use, I think the nylon-insert type is both necessary and sufficient.

To stop a standard hex nut from coming loose, you can put a drop of **thread locker**, such as **Loctite**, on the bolt before you tighten the nut. Loctite is similar to Krazy Glue. It sets quickly, and glues the nut in place. Some variants of Loctite grip less permanently than others, allowing nuts to be removed if sufficient force is applied.

Turning a Nut

The adjustable wrench that I suggested in this project is not a very satisfying tool to use, because of its tendency to loosen and slip. In doing so, it can round the corners of the nut. A severely rounded nut can be difficult to remove in the future.

Socket wrenches grip all six corners of a nut, and are often sold in sets. Typically the set consists of multiple sockets and a ratchet that plugs into the socket of your choice. People who work on cars typically have large arrays of sockets, in both US and metric sizes. If you just want to be able to deal with nuts and bolts once in a while, you can get a miniature set of sockets very cheaply, like the one in Figure 14-39.

You can also get sockets such as those shown in Figure 14-40. They can be used in a hand driver that grips the shaft of each socket. They can also be used in an electric drill, although many sets are not rated for power tools.

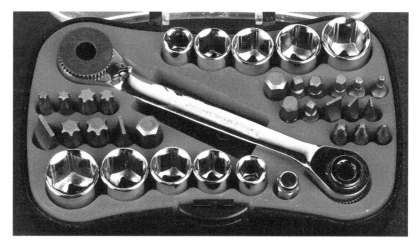

Figure 14-39. A miniature socket-wrench set with the most popular metric and US sizes, and assorted bits for various head configurations. The arm terminates in a ratchet that connects with each socket via a detachable plug.

Nut drivers are like screwdrivers, but with little sockets at the end. A set of nutdrivers is shown in Figure 14-41. They are very convenient, but the amount of turning force that you can apply is smaller than the force that a socket wrench can deliver.

Figure 14-40. Socket bits for an electric drill.

A problem with sockets is that they can't screw the nut very far up the thread of a bolt, because the end of the bolt hits the top of the socket. You can, of course, get deep-well sockets, but you'll end up paying more for them.

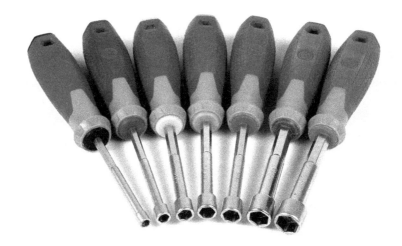

Figure 14-41. A set of seven nut drivers.

Bending It

At this point I want to introduce you to the world of plastics. Depending what type of plastic you use, it can be easier to work with than wood, and doesn't necessarily cost more. Plastics usually don't require a coat of paint or polyurethane when the job is done, and they can enable creative designs that would be difficult or impossible with other materials.

I'm going to begin with **ABS**, because it's versatile and affordable. You may not have heard of it, but everyone comes into contact with it at some time or other. Lego blocks and other children's toys are made of ABS. Many 3D printers work with ABS. I am fairly sure that the wheeled trash can outside my house is molded from ABS, and so far as I can tell, even the little computer speakers that I have on my desk are in ABS enclosures.

There are so many things you can make very simply by bending plastic, I'm going to include four mini-projects here that only entail cutting and bending. Gluing and screwing parts together will be dealt with in the next two chapters, and transparency and colors in the two chapters after that.

Getting Acquainted With ABS

ABS is an acronym **acrylonitrile-butadiene-styrene**. The "acrylo" part of its name tells you that chemically speaking, ABS has some family ties to **acrylic** plastics that go by brand names such as **Lucite**, **Plexiglas**, or **Perspex**.

It looks and feels quite different, though. Acrylics are transparent, brittle, and relatively hard. ABS is most often sold in opaque black and white sheets. It can be bent or curved at a relatively low and safe temperature, doesn't crack or split unless severely abused, and while being soft enough to be shaped easily, it is tough enough to take substantial loads.

You can buy sheets online in sizes from 12" x 12" upward. (Just check eBay.) If you live in a city, you can probably find a retail source for sheets that are 48" x 96". Thicknesses include 1/16", 1/8", 3/16", and 1/4".

Preparation

Your first step with a new piece of ABS is to clean it. This will be much easier than after you have cut it into pieces. Wipe both sides with a damp, nonabrasive rag or sponge, adding a trace of dishwashing liquid if necessary.

NEW TOPICS IN THIS CHAPTER

- Cutting and bending ABS plastic

See next page for a materials list.

YOU WILL NEED

- Two-by-four pine, any condition, length 12"
- One-by-six pine, a few knots, not warped, length 12"
- One-by-six pine, no knots, not warped, length 12"
- Plastic sheet, white ABS, ⅛" thick, 12" wide, length 24"
- Ceramic tiles, 4" x 12" minimum, quantity 3
- Bolts, ¼" x 1", 20 tpi, hex or Phillips pan head, quantity 2
- Locknuts, nylon inserts, ¼", 20 tpi, quantity 2
- Fender washers, ¼" center hole, 1" diameter, quantity 4
- Heat gun, small size, 350W approximately
- Heat-resistant cotton gloves (optional)
- Deburring tool (optional)

Also, as listed previously: Tenon saw, miter box, trigger clamps, ruler, speed square, rubber sanding block, work gloves, awl, pliers, utility knife, adjustable wrench, electric drill and drill bits, countersink, dust mask (optional), safety glasses (optional), plywood work surface, and sandpaper.

Check the Buying Guide on page 248 for information about buying these items.

When handling ABS, always remember that you probably won't be painting or otherwise improving its finish. It scratches easily, so treat it gently. Avoid laying tools on it, and when you place it flat on a work surface, make sure there are no sharp objects such as nails or screws underneath it.

Most ABS is textured on one side and smooth on the other. The textured side is slightly tougher and doesn't show scratches so easily, and I think it looks nicer. I'm assuming this will be the outside surface for your project.

To draw a pattern on the smooth side of ABS, you can make pencil lines, but they will be very faint. Personally I like to use a fine-point water-based roller-ball pen. This makes cutting easier because the lines are so visible—assuming you are not using black ABS, of course. If you're wondering how to mark that, stretching pieces of masking tape is an option.

You should be able to wipe away water-based ink without any difficulty, but don't use a permanent marker. The most effective solvents for removing marker ink also dissolve ABS.

Cutting ABS

Most everyday woodworking tools can be used with ABS, although a saw that does not have hardened teeth may be dulled more rapidly than if you are cutting wood.

Power saws are a problem. The heat generated by friction causes ABS to melt in contact with a circular saw, leaving smears of plastic on the sides of the blade. These smears melt again when you are cutting something new, and can be sticky enough to grab the material and throw it at you with great force.

Plastic-cutting blades are available, made of thicker steel that absorbs more heat. Blade lubricant can also help, but ABS is so soft, I use a handsaw even for cutting the full length of a 96" sheet.

Unlike plywood, ABS doesn't splinter, and a tenon saw with hardened teeth will make a fairly clean cut. A handy item known as a **deburring tool** can smooth ragged edges while adding a narrow bevel, just with a single pass. This is much more efficient and satisfactory than using sandpaper.

The tool is shown in Figure 15-1. It has a sharp edge (you can see it reflecting the light in the photograph), but this terminates in a rounded tip that makes the tool quite safe to use. Wearing gloves is still a good idea, just in case.

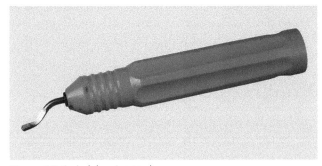

Figure 15-1. A deburring tool.

The way in which a deburring tool removes a thin ribbon of plastic from the edge of ABS is shown in Figure 15-2.

Heating ABS

To bend or curve a sheet of ABS, you have to raise its temperature to 105 degrees Celsius or about 220 degrees Fahrenheit, just above the boiling point of water. There are two ways to do this.

The easiest way is the expensive way, using a **plastic bender** consisting of an electrical element in a long, ceramic channel. An example is shown in Figure 15-3. You plug it in, wait about 10 minutes for it to get hot, and then lay a piece of plastic over it. The element heats the plastic along the line where you want the bend. After 30 to 60 seconds (depending on the thickness of the plastic), you can bend it with reasonable precision—usually within ⅟₁₆".

Unfortunately, this type of plastic bender is not cheap. If you go online, you can find plans and videos suggesting how to make your own version, but that kind of DIY project, using a hot element powered with house current, seems a bit iffy to me.

I think you should use a **heat gun**. This is not only less expensive, but can make multiple bends that would get in the way of each other on a plastic bender.

A heat gun is like a hair dryer, but more powerful. (A hair dryer is not hot enough to soften ABS.) In addition to bending plastic, it has other uses. If you are doing hobby-electronics, **heat-shrink tubing** is designed to shrink when you apply a heat gun. Sticky labels will unstick themselves more easily if you blow hot air at them, and if you have a cold workshop, epoxy glue flows more easily and sets more quickly when you apply some heat. Quite often, I find myself thinking, "Maybe a heat gun can help with this little problem."

Figure 15-2. Using a deburring tool to smooth and bevel an edge.

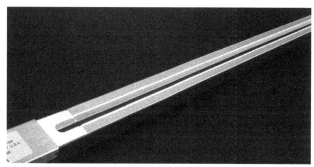
Figure 15-3. This plastic bender uses an electric heating element in a ceramic channel.

Figure 15-4. A small heat gun.

You can buy either a full-size heat gun, using about 1,000 watts, or a small-size version that consumes approximately 350 watts. For our purposes, the smaller, less expensive type will be sufficient. An example is shown in Figure 15-4. It costs less than the larger type, and is safer to use—so long as you use it sensibly.

Caution: Burn Risk

At the risk of emphasizing the obvious, heat guns do get hot. Don't point one at yourself, or at other people. Don't put your hand close in front of it to check that it's working, and don't lay it down and walk away from it while it's running. Also, a heat gun should not be used with materials that burn easily, or if there is flammable vapor in the air.

If you are wearing 100 percent synthetic clothing such as polyester, nylon, or spandex, there is the potential risk of melting the cloth if you point the heat gun toward yourself accidentally at close range.

Higher-powered heat guns often incorporate a steel tube to direct the heat, and this tube will remain hot enough to burn your hand for a while after you switch off the power. Therefore, don't leave a heat gun lying around where children may pick it up or pets may be hurt by it. And unlike the character in Figure 15-5, you should wear heat-resistant gloves while bending plastic. Pure cotton gloves or oven mitts are acceptable.

Figure 15-5. An unlikely scenario, but still, maybe wearing heat-resistant gloves would be a good idea.

Focus the Heat

When hot air emerges from a heat gun, it tends to spread out. This is okay if you want to make a gentle curve in the plastic, but if you are hoping for a sharply defined bend, such as the corner of a box, you will need a way to confine the heat to a narrow strip.

For this purpose, you could use small pieces of plywood as heat shields. However, a heat gun is hot enough to scorch wood, and can even make wood burn if you run it for long enough at short range.

I think a better idea is to use a nonflammable material, and **ceramic floor tiles** are a good choice. You can buy them at any big-box hardware store, and three of them will be sufficient.

A Test Bend

To see how easy it is to bend ABS, cut a piece measuring about 4" x 2" and place it between two tiles on your work area, with the end sticking out, as in Figure 15-6.

Figure 15-6. ABS plastic ready for bending.

Hold the third tile to cover the part of the plastic that you want to protect from the heat, and use the heat gun on the narrow strip that is exposed, as in Figure 15-7.

Most heat guns have two settings. If you are using a small heat gun, use the high setting; if you are using a large heat gun, use the low setting. Keep the tip of it 2" away from the exposed ABS while you direct the hot air up and down the strip in a steady rhythm. After about a minute, when you press on the plastic, you should find that it isn't as springy as before. Switch off the heat gun, and see if the plastic will bend as in Figure 15-8. If it resists, heat it a little more.

You don't necessarily have to bend the plastic while it is clamped between the tiles. You can take it out and bend it in your gloved hands, and you should still get a well-defined folded shape. This is because while you were using the heat gun, the tiles not only blocked the heat but absorbed some of it.

Figures 15-9 and 15-10 show the concept. If you heat plastic that is unprotected, the plastic conducts some of the heat along its length, so the hot area is relatively wide. If you clamp the plastic between tiles, they conduct heat away, so that only the area under the heat gun gets hot enough to bend.

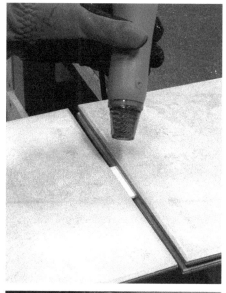

Figure 15-7. Heat is focused on a strip ½" wide, or less.

Figure 15-8. A simple bend.

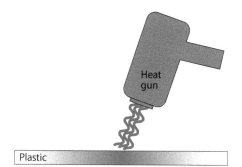

Figure 15-9. Without any protection, plastic becomes hot over a wide area.

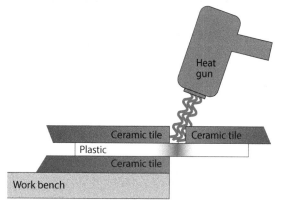

Figure 15-10. Ceramic tiles confine the heat and conduct it away from a narrow area.

How long should the heating process take? This depends on several factors.

- Thin plastic heats much more quickly than thick plastic.
- A wider or longer exposed strip of plastic takes more time, because a fixed amount of heat is spread over a larger area.
- The distance of the heat gun from the plastic will make a difference.
- If you bend multiple pieces of plastic, each one may take less time than the last, because the tiles become hotter. The heat gun also becomes hotter with use.

If you overheat ABS, you will scorch it. First it turns yellow, and then brown. This change is not reversible. If your plastic becomes soft in less than 30 seconds while using a small heat gun, you're probably making the plastic too hot.

When making a bend, always remember to curve the plastic *away from the hot side*, so that the hot surface stretches around the outside of the bend. If you try to do it the other way, the hot surface may bunch up or create an ugly fold inside the bend.

After you make a bend, use a speed square to check the angle, and when it looks the way you want it, you can spray it with a thin mist of water or wipe it with a wet sponge. It should stiffen almost immediately.

You may be tempted to use a pencil or pen to indicate where you want a bend to be. This is not a good idea, because the heat can make the mark permanent. I generally mark a bend with a tiny dot on the edge of the plastic. This can be scraped off later with a utility knife.

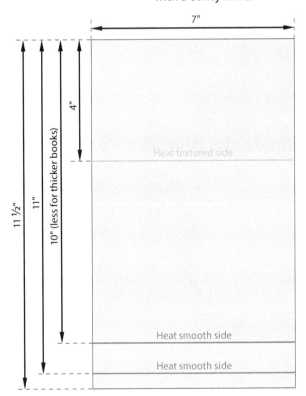

Figure 15-11. The plan for a small paperback book stand.

If you're using large tiles, and the one that you're holding is too big and heavy, you can break it into smaller pieces. You can try to control the break by scoring the tile with a glass cutter, if you have one. Either way, you can snap the tile across a length of two-by-four. If you are reluctant to do this with your hands, and you have a floor that won't be easily damaged, hold down one end of the tile with your foot, and step on the other end.

Be careful with broken edges, which can be sharp.

Paperback Book Stand

I'm starting with this project because it's the simplest. Just three folds will do it.

If you like to sit and read a book without having to hold it up, a book stand solves this problem. You can buy one, of course, but wouldn't you rather make one that is exactly the right size for your needs? If you only read paperbacks, the stand can be compact. If you prefer art books, it can be big. I'll describe a compact one, leaving you to change the scale if necessary.

The plan is shown in Figure 15-11. I'm guessing that you have a piece of ABS 12" wide, because this is the standard minimum width when you buy it online. The plan is drawn with this in mind, as it requires a piece of plastic 11½" x 7".

When you receive a sheet of ABS, it may have a sawn edge, which is likely to be straight but rough. If it has a "natural" edge, it can look like the closeup in Figure 15-12. Either way, you're going to have to cut your own reference edge, which will be both straight and smooth. This is similar to the procedure that you followed when using plywood in Chapter 10.

To create your reference edge, I suggest you cut along a piece of guide wood as shown in Figure 15-13. A sacrificial piece under the plastic isn't necessary, because ABS doesn't splinter, and a tenon saw with hardened teeth will make a fairly clean cut.

The two-by-four under the plastic is only there to provide some support beyond the work area. A ⅛" thick sheet of ABS is not very rigid, so it needs support when you are cutting it. Sawing it closer to the edge of your work area is not an option, because the clamps would get in the way.

Saw gently, being careful to keep the side of the saw vertical. When you complete the cut, you can clean the underside with a deburring tool or a few strokes of a sanding block.

Now use your speed square to draw a line at 90 degrees from your reference edge, as in Figure 15-14. Measure along this line and make a mark at 7". I'll call it Point A.

Measure 11½" along the reference edge, as in Figure 15-15, bearing in mind that the piece you need to make your paperback book rest is 11½" long.

Figure 15-12. The "natural" edge of a piece of ABS can look like this.

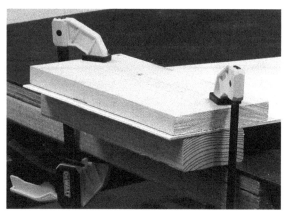

Figure 15-13. Setup for cutting a reference edge.

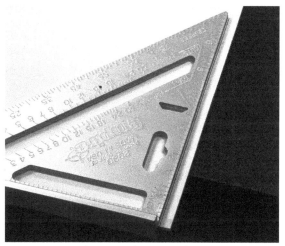

Figure 15-14. A line at 90 degrees to the reference edge.

Figure 15-15. Measure 11½" along the reference edge.

Use your speed square to draw another 90-degree line at the 11½" mark, as in Figure 15-16. Extend this line and mark it 7" from the reference edge. I'll call this Point B. Now you can draw a line connecting Point A and Point B, and it will be parallel with your reference edge, and the same length. You can cut along this line, as in Figure 15-17, and then cut the two other edges of the rectangle to its finished size. (The saw happened to be held at this angle when the photograph was taken, but should be used at a shallow angle while making the cut.)

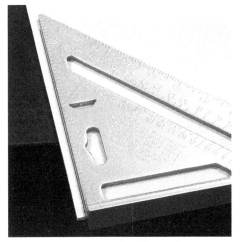

Figure 15-16. A second 90-degree line.

Figure 15-17. Cutting out the rectangle.

Now you need to make marks on the edges to define where the bends will be, as shown in the plan in Figure 15-11. The easiest way to do this is to clamp the ABS against a piece of two-by-four, so that you have something to rest the ruler on, as in Figure 15-18.

After you make marks along one edge, you can use your speed square to copy them across to the other edge, as in Figure 15-19.

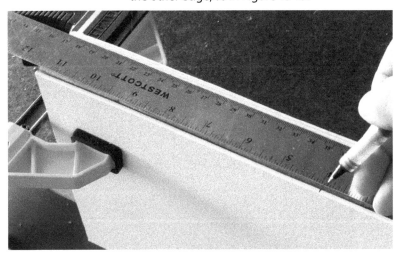

Figure 15-18. Marking where the bends will be.

Figure 15-19. Copying the bend marks to the opposite edge.

You're ready to start bending. The plan in Figure 15-11 assumes you'll want the textured side of the ABS on the exposed face of the book stand. Don't forget: you always bend the plastic away from the heat, because the hot side has to stretch around the bend. The outside of the bend is a longer distance than the inside.

Beginning at the top and working downward, your first bend, at the 4" mark, requires you to heat the textured side. So, the plastic goes between the tiles textured-side-up. You should be able to see the edge marks about ⅛" beyond the tiles. Check that the plastic is at 90 degrees to the tile edge by using your speed square, as in Figure 15-20.

Apply the heat gun as in Figure 15-21. When the plastic has softened, remove it quickly, hold it between your two hands, and bend it along the line that you heated till it makes a sharp angle as in Figure 15-22. To see why the sharp angle is necessary, look ahead to the finished book stand in Figure 15-26.

For the next bend, the plastic goes into the tiles smooth-side-up, as in Figure 15-23. Before you apply the heat gun, add a piece of tile to protect the exposed area beyond the bend.

When the plastic is sufficiently soft, your second bend is at 90 degrees, as shown in Figure 15-24.

For the third bend, the plastic goes into the tiles smooth-side-up, as shown in Figure 15-25. Only ½" of the plastic remains inside the tiles,

Figure 15-20. Check before bending!

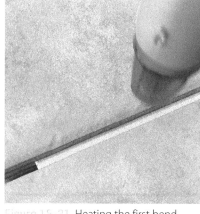

Figure 15-21. Heating the first bend.

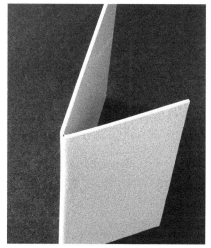

Figure 15-22. The first bend is a sharp one.

Figure 15-23. Ready for the second bend.

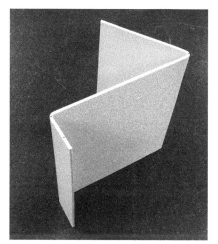

Figure 15-24. The second bend, completed.

Figure 15-25. Ready for the third bend.

Figure 15-26. Job done.

so you'll need to clamp it firmly. A short bend is always more difficult than a long one.

After the third bend—that's it!

Use sandpaper to round the four corners, and the job is done. The finished paperback book stand is shown in Figure 15-26.

Now consider how you would have made this out of plywood. Probably you could have used two triangular side pieces, to support the slanting book rest.

But how would you have made the lip at the bottom? I don't think it would have been quick or easy, and the end product would have weighed more while probably not looking as nice.

So what else can we make with ABS?

Adjustable Paper Towel Dispenser

I decided that I wanted a paper towel dispenser which will allow me to adjust its turning resistance. I think this concept is useful, because the towel roll should be loose enough to turn, but stiff enough to resist when you tear a towel along its perforations.

Figure 15-27. A towel dispenser that allows you to adjust the turning resistance.

The design in Figure 15-27 consists of a simple bracket to mount on the wall, and two inserts that push into the cardboard tube in the center of a roll of towels. You have to squeeze the lugs that stick out of the inserts when you push them into the tube. Once inside the roll, the lugs are springy enough to grip it. You adjust the turning resistance by tightening or loosening a ¼" bolt at each end of the roll holder.

The plan for the wall bracket is shown in Figure 15-28. This is longer than the likely 12" width of your ABS sheet, unless you decide to make it in two pieces. I'll leave that to you. It's a very simple design, requiring just two bends. I think you can make those without any need for photographs, at this point. The holes are centered vertically between the two long edges.

A plan for each end piece is shown in Figure 15-29. Saw cuts are in red. The circles indicate holes that you drill through the sheet. I put a cross in the center of each one so that if you enlarge and print this plan, you can easily prick through the center of each circle into the plastic.

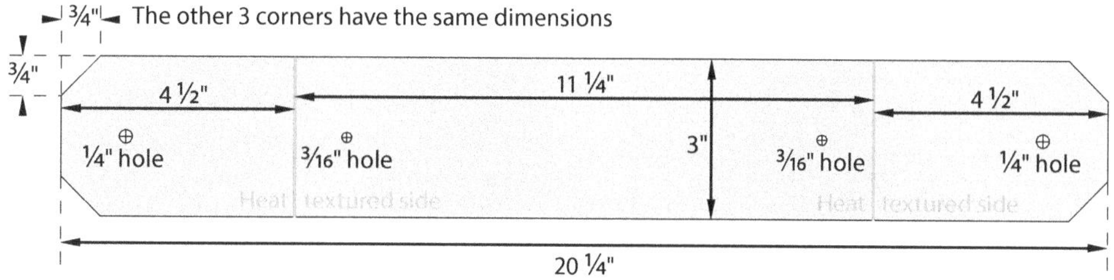

Making the ¼" hole in the center shouldn't be a problem, but a ⅜" drill bit grabs the plastic, digs in, and can make a mess. You can minimize the problem by countersinking to the limit of the holes established by the straight lines, as in Figure 15-30. Then use a ⁵⁄₁₆" drill bit, and finally the ⅜" bit. Make sure the plastic is firmly clamped, and run the drill as slowly as possible.

You may wonder why the ⅜" holes are necessary. The reason is that whenever you have two saw cuts meeting inside a piece, you should always drill a circular hole first, and then saw into the holes, as shown in Figure 15-31. If you allow two saw cuts to meet at a sharp point, and a bend line is nearby, the plastic will tend to split where the cuts meet. The curve at the inside of the hole allows the plastic to stretch instead of splitting.

You have to angle the saw up till it is almost vertical as it approaches each of the holes. Proceed cautiously with

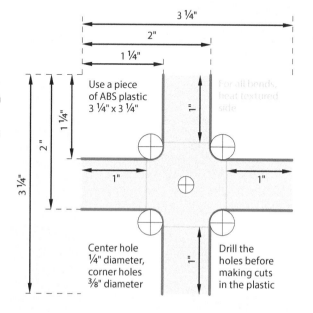

Figure 15-28 The plan for the bracket for the paper towel dispenser.

Figure 15-29 An end piece for the paper towel dispenser. Make two copies.

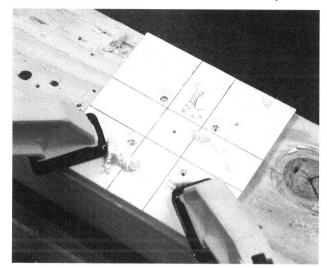

Figure 15-30 Preparing to drill ⅜" holes in an end piece.

Figure 15-31 Sawing into one of the ⅝" holes.

Figure 15-32. Cleaning away some plastic remnants with a utility knife.

Figure 15-33. Bending lugs on an end piece.

Figure 15-34. Testing an end piece.

Figure 15-35. An end piece mounted in the bracket.

the last few millimeters of the cut. If you stop just before the hole, ABS is soft enough to let you wiggle or twist the piece away. Any remnants can be sanded or pared away with a utility knife, so long as you cut down into some scrap wood, as shown in Figure 15-32. Don't pull the knife toward you.

Bending the lugs on the end pieces is very straightforward. All the bends can be smooth-side-up. Figure 15-33 shows one of the end pieces with three of the lugs bent, and one remaining.

Test your first end piece to make sure it's a tight fit in a roll of towels, as in Figure 15-34. You can always reheat the plastic to bend the lugs in a bit or out a bit.

Use a 1" long, ¼"-size bolt, two 1" fender washers, and a locknut to mount each end piece in the bracket that holds the towels, as in Figure 15-35. The threads on the nut and the bolt must match (usually, 20 per inch). If you built the pantograph in Chapter 14, these items will be familiar to you. Tighten the bolt to increase resistance to rotation.

The towels in the dispenser are shown in Figure 15-36. I happened to have flat-headed ¼" bolts, so I countersunk them into the bracket for a cleaner look. Pan-headed bolts would do just as well.

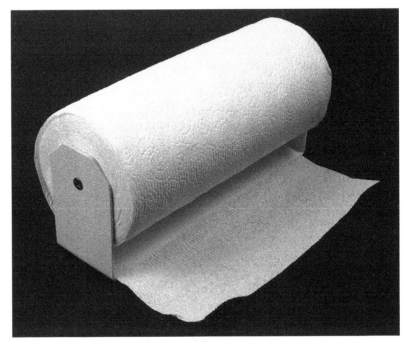

Figure 15-36. Towels ready for optimal dispensation.

Shower Caddy

Maybe you already have one of these. I'm talking about a little rack or basket that contains shampoo and other essentials, hanging from the shower head in your bathroom. Even if you have one, it may not be the right size or shape for the bottles that you want to put in it. This is a good reason for designing your own, as part of the pleasure of making things is that you can get exactly what you want.

If you make a smaller version, it can be a soap dish that hangs from a suction cup on the tiling around your shower. My design in Figure 15-37 has no dimensions on it, because you can adjust the scale to suit your needs.

The first step in designing your own folded ABS object is always to draw the plan on a piece of paper, cut it out, fold it, and see if it looks good. Once you have it the way you want it, tape the paper to the smooth side of the ABS and draw around it. The drawing doesn't have to be super-neat. My version, in Figure 15-38, certainly wasn't.

After drilling holes where cuts will meet, you can cut around the outline using a tenon saw for straight edges and a coping saw for curved edges.

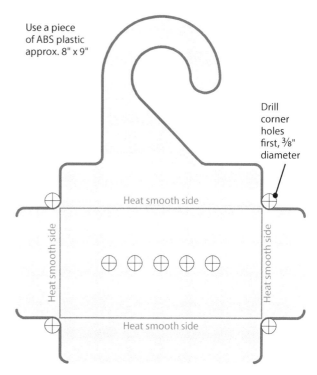

Figure 15-37. A shower caddy or soap dish, depending on the size you choose to make it.

This is a simple object to bend, because all the bends are in the same direction, and there are only four of them. I don't think you'll need step-by-step photographs, so I'll just show the finished object in Figure 15-39.

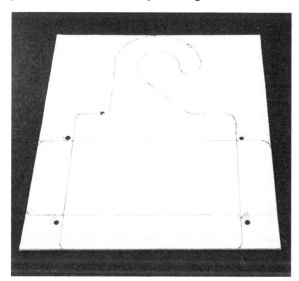

Figure 15-38. Rough outline for the shower caddy, sketched on the smooth side of ABS, with the holes drilled.

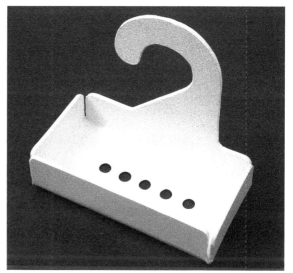

Figure 15-39. This version is soap-dish sized, but making a larger version would be just as easy.

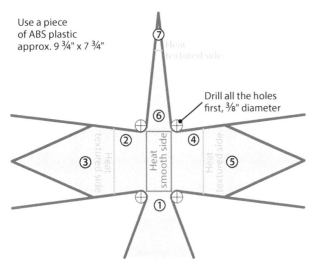

Use a piece of ABS plastic approx. 9 ¾" x 7 ¾"

Drill all the holes first, ⅜" diameter

⑦ Heat textured side

⑥

② Heat textured side

③ Heat textured side

Heat smooth side

④

⑤ Heat textured side

①

Figure 15-40. The numbers tell you the necessary sequence for the bends.

Figure 15-41. Using vertical pieces of tile to protect previous bends from the heat.

Origami Crane

This is a slightly more difficult fabrication project because it mixes bends in opposite directions. After each bend becomes rigid, it tends to get in the way of the next bend.

Be careful to make the bends in the correct sequence. The numbers in the plan tell you what the correct sequence is. See Figure 15-40.

After you cut and drill the shape in the same way as in previous projects, start the bending sequence with the tail. By the time you get to the second wing (bend number 5), you'll have to be creative to avoid reheating previous bends by using vertical pieces of tile, as shown in Figure 15-41.

The finished crane is in Figure 15-42. If you like the look of it, why not fold a couple more, and use them in a mobile?

I think you can see from this chapter that a bendable material opens up some fascinating design possibilities. But glue and screws also have their uses, as I'll show you in the next two chapters. After that, I'll describe a concept that applies glue, screws, *and* bending to plastic of two different types.

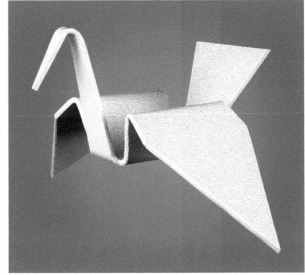

Figure 15-42. The finished crane, folded from ABS, resembles traditional Japanese paper origami.

Chapter 16

A Better Box

In Chapter 10 I promised to build a better box, and that time has come. In fact, I am going to suggest a pair of better boxes. The first will be in this chapter, and the second will be in the next.

Making the Pieces

The box that I am hoping you will build is shown in a 3D rendering in Figure 16-1. All of the joints will be glued, using an ABS solvent that you can buy from plastic supply sources.

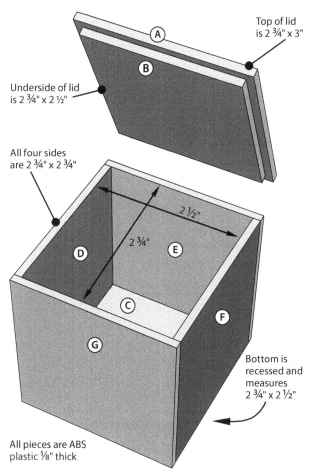

Top of lid is 2 ¾" x 3"

Underside of lid is 2 ¾" x 2 ½"

All four sides are 2 ¾" x 2 ¾"

2 ½"

2 ¾"

Bottom is recessed and measures 2 ¾" x 2 ½"

All pieces are ABS plastic ⅛" thick

Figure 16-1. How the pieces of your box will fit together.

The pieces that you need for the first box are shown in Figure 16-2. I chose the dimensions so that everything can be taken from two strips of plastic, each measuring 12" x 2¾". All four sides of the box are square, so you don't have to worry about which way up the pieces should be. (Because two of the sides overlap the edges of the other two sides, the box itself is not quite square.)

You can cut the strips in the same way that you cut pieces of ABS in the previous chapter. Establish a reference edge, if necessary, then make the strips by cutting parallel with it. After you have the strips, you can set up your miter box and use it to make

NEW TOPICS IN THIS CHAPTER

- Using solvent cement with ABS plastic

YOU WILL NEED

- Two-by-four pine, any condition, length 12"
- One-by-six pine, a few knots, not warped, length 12"
- Square dowel, ¾" x ¾", length 3"
- Plastic sheet, white ABS, ⅛" thick, 12" wide, length 12"
- Solvent cement, compatible with ABS, acrylic, and polycarbonate, 4 ounces
- Applicator for solvent cement, plastic bottle with hollow needle
- Disposable chemical-resistant gloves, nitrile preferred
- Goggles, chemical-resistant
- Clean rags, size 12" x 48" approximately
- Cardboard, any thickness, size 12" x 12" approximately

Also, as listed previously: Tenon saw, miter box, trigger clamps, ruler, speed square, rubber sanding block, work gloves, awl, masking tape, utility knife, deburring tool (optional), dust mask (optional), plywood work surface, and sandpaper.

Check the Buying Guide on page 248 for information about buying these items.

Figure 16-2.
The pieces that you will need to cut.

These pieces can be cut from strips 2 ¾" wide.

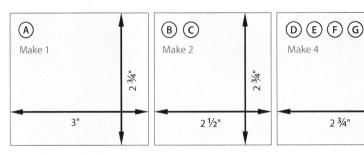

(A) Make 1 2 ¾" 3"

(B) (C) Make 2 2 ¾" 2 ½"

(D) (E) (F) (G) Make 4 2 ¾" 2 ¾"

Figure 16-3. Tooth marks from a saw on a freshly cut edge.

Figure 16-4. Sanding the edge of a square.

Figure 16-5. The saw-tooth marks have been sanded away. Some dust from the sanding is still visible, and should be wiped off before applying solvent cement.

the pieces shown in Figure 16-2. Remember to measure one piece, then cut it, then measure the next piece, and cut it, just as you did in Chapter 1 with the dowel. The dimensions have to be precise, so don't ignore the thickness of the saw blade.

Solvent Basics

The type of adhesive that is most often used with plastic is a **solvent cement**. While ordinary glue forms a layer between two objects, a solvent dissolves plastic so that the objects melt together. The solvent then gradually disappears through evaporation.

Solvent cannot fill gaps, so when you are making a butt joint where the edge of one piece of plastic is glued to the surface of another, your sawn edge has to be straight and smooth. This requirement is a bit more demanding than anything we have dealt with before, and a deburring tool isn't really appropriate, because it bevels the edge, leaving less of it available for contact with the other piece.

A typical fresh-sawn edge of a ⅛" sheet of ABS is shown in Figure 16-3. You can see the marks made by the saw teeth. Clamp some 80-grit sandpaper to a rigid, flat, horizontal surface and rub the edges of your sawn pieces over it, as shown in Figure 16-4, taking care to keep each piece level while holding it vertically, so that the edge doesn't become rounded. With a small amount of sanding, your sawn edge should look like the one in Figure 16-5.

The solvent that you are going to use will work with ABS even though the word "acrylic" is on the can. It is shown in Figure 16-6, alongside an applicator that consists of a squeeze-bottle with a hollow needle. The needle is large and blunt, and should not be hazardous to use, if you are reasonably careful. The solvent itself is another matter.

Solvent Hazard Warning

The ABS solvent that I am recommending looks like water, but is very volatile and gives off fumes that you don't want to breathe too much. Try to avoid getting it on your hands, and definitely keep it out of your eyes. Wear chemical-resistant gloves and goggles, and maintain good, active ventilation. Also, avoid using the solvent near any naked flame. It definitely requires a nonsmoking area.

The dispenser can squirt droplets unpredictably. If you merely drop it on your work area, the impact can provoke a little squirt of solvent. You need to take this risk seriously.

If you wear eyeglasses that have polycarbonate lenses, this solvent will dissolve them. Therefore, you should use goggles in addition to any eyeglasses that you wear.

Read the instructions on the can, and if there is any conflict between them and my instructions, the can takes precedence.

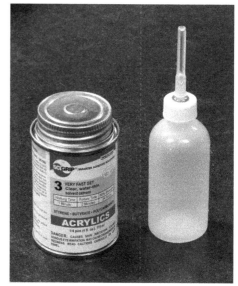

Figure 16-6. Solvent is supplied in a can, and must be transferred to a squeeze-bottle with a blunt, hollow needle for application.

Solvent Application

You may want to protect your work area by laying cardboard over it before you begin. Then lay four or more thicknesses of clean rags to absorb any solvent that splashes or escapes from underneath a joint.

Opening a can of solvent can be difficult, because it may be sealed for shipment. This means that when you remove the screw-cap, you will find a metal seal underneath that is not removable. If you have an old-fashioned can puncher, you can use that to make a hole in the seal. Alternatively, use an awl to punch a hole. Tap the awl gently with a hammer, and be careful to avoid tipping the can over.

After you have perforated the metal seal, you'll have to store the can with the cap screwed back on tightly to prevent evaporation and protect you from breathing the fumes.

The squeeze-bottle with its hollow needle is cheaply available from plastic suppliers online. Transfer solvent into the bottle by squeezing air out of it and inserting the needle into the can, so that it can suck up some liquid. The bottle only needs to be about one-quarter full.

When you are not using the bottle, keep the little plastic sleeve over the needle to prevent solvent from evaporating through it.

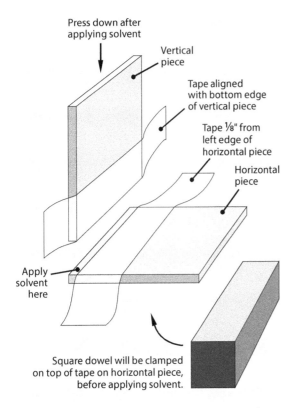

Press down after
applying solvent

Vertical
piece

Tape aligned
with bottom edge
of vertical piece

Tape ⅛" from
left edge of
horizontal piece

Horizontal
piece

Apply
solvent
here

Square dowel will be clamped
on top of tape on horizontal piece,
before applying solvent.

Figure 16-7. Setting up two pieces for application of solvent cement.

Figure 16-8. Ready for cementing.

Demonstration videos usually show someone joining two pieces of transparent acrylic. The pieces are held in position while solvent is applied at the joint, and you can see the solvent being sucked into the joint by surface tension.

Your situation is more difficult, as you can't see what's going on when you are using opaque ABS. Here's the procedure that I suggest.

Start with the square sides of the box, labeled D, E, F, and G in Figure 16-2. Lay one square of ABS flat, with the textured side down. I'll call this the horizontal piece. The other square will stand on it. I'll call this the vertical piece. The solvent will bond the edge of the vertical piece to the surface of the horizontal piece, to make a butt joint. A piece of square dowel will help to make the joint 90 degrees.

Remove any moisture or grease from the surfaces that you will be gluing by rubbing them with a dry cloth or paper towel. Then apply a strip of Scotch tape to the vertical piece, on its smooth side, aligned with the bottom edge, and a strip of tape parallel with the edge of the horizontal piece of ABS, but ⅛" away from it. The ⅛" margin is where the solvent should go. Figure 16-7 shows the plan.

The tape will provide protection from stray drops of solvent. I suggest genuine 3M Scotch tape that has "transparent tape" printed on its red label. "Magic tape" is more difficult to remove.

A piece of square dowel will help you to make a 90-degree joint. Place it on the horizontal piece, aligned with the edge of the tape, and add a clamp. The setup is shown in Figure 16-8.

Now for the tricky part. I am assuming you have some solvent in your squeeze bottle, and you have removed the protective sleeve from the needle. Hold the solvent bottle vertically, and squeeze air out of it. Then release your squeezing pressure while, at the same time, you turn the bottle over. The bottle will draw air in through the needle, which should stop solvent from running out.

If you don't do this, solvent will dribble out of the needle rapidly and unpredictably, making a mess. Practice inverting the bottle a couple of times in an area where nothing will be damaged if solvent escapes.

When you're ready to glue the two pieces, invert the bottle and release just enough pressure to allow a little solvent to come out, while you run the end of the needle along the margin between the dowel and the edge of the plastic. This has to be done quickly. The position of the needle is shown in Figure 16-9.

Quickly put the squeeze-bottle down and place the edge of the vertical square of plastic on the wet solvent, with the textured side facing outward. Press down on it hard, while also holding it against the dowel to keep it in alignment.

Figure 16-9. How to position the syringe. See text for full details.

Figure 16-10. A completed joint.

The solvent creates a bond almost immediately. This is a relatively weak bond, but good enough for you to remove the clamp and the dowel. Pick up the pieces, being careful not to disturb their alignment. You need to get them off the rags in case some solvent has trickled under the horizontal piece, where it will dissolve the plastic into the rags.

After a minute, any stray solvent will have evaporated, and you can put the pieces down. Check that they are still at 90 degrees. You have about 30 seconds in which to adjust their angle before the joint starts to harden. A successful joint is shown in Figure 16-10, still with a piece of tape attached to the vertical piece.

Solvent requires about 24 hours to evaporate completely from inside the joint. You can continue working with the pieces in the meantime, so long as you are very gentle with them.

Additional Joints

Make another butt joint, identical to the first one, with your two remaining squares of ABS. Now you have two L-shaped combinations of pieces. When you put them together, they will form the front, back, and sides of your box. Figure 16-11 reminds you that there is a right way and a wrong way to do this.

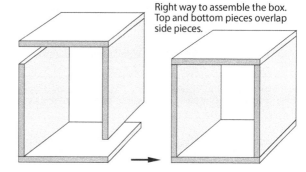

Right way to assemble the box. Top and bottom pieces overlap side pieces.

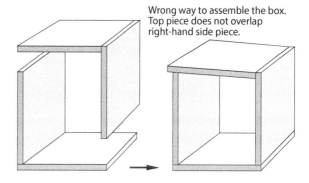

Wrong way to assemble the box. Top piece does not overlap right-hand side piece.

Figure 16-11. Assembly of two pairs of side pieces.

Figure 16-12. Allow solvent to flow into the cracks between clamped sides of the box.

Figure 16-13. The base will have to be sanded to fit. (In this photo, the box is upside-down.)

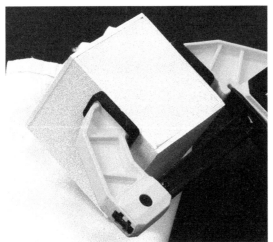

Figure 16-14. The base now fits the bottom of the box, and can be clamped in place.

Clamp the two L-shaped pieces in position, without using any solvent, yet. To avoid stressing the joints, use minimal force on the clamp—barely enough to hold the pieces in position. Now you can use the needle to trickle solvent into the crack inside between two of the pieces that are not glued yet, as in Figure 16-12. It should flow into the crack very rapidly. Turn the pieces over and glue the remaining pair in the same way. After a minute you can release the clamp.

Let the joints set for an hour or so. Now you can use one of the pieces measuring 2¾" x 2½" to form the bottom of the box. (This is piece C in Figure 16-1.) Probably, it won't quite fit, and will look like the one in Figure 16-13. Sand it repeatedly until it just slips into place. Don't force it in—you'll spring open the joints that you made previously.

When the base is the right size, slip it into place and clamp the box gently, so that it holds the base, as shown in Figure 16-14. Turn the box over as in Figure 16-15 (base-down), and you can use the needle to apply solvent around the edges from the inside. Once again, some solvent may trickle through the cracks and drip from the bottom. Pick the box up to allow any stray solvent to evaporate.

You may still see little gaps between the base and the box, which the solvent cannot fill. But probably you have enough points of contact to make a reasonably secure bond.

Now for the lid. In Chapter 10, the wooden blocks inside the box had a useful function: they stopped the lid from dropping down inside. A different arrangement is needed for this box.

The rendering in Figure 16-1 shows what I have in mind. The lid will have two layers. The top layer, identified as A in the figure, will overlap the sides of the box, so it won't fall in. The lower layer of the lid, identified as B, will be about the same size as the inside of the box, functioning to center the lid.

Start by sanding the second layer till it fits easily. Then place the top side of the lid (piece A) textured-side-down on the rags. Make sure it is horizontal, and add a couple of drops of solvent in the center. (If the lid is not horizontal, the solvent will run off it.) Immediately take the underside of the lid (piece B) and position it textured-side-up on piece A, with an equal margin all around the edges. Press firmly, and it should stick immediately.

The finished box is in Figure 16-16, with the lid removed and turned upside-down, and in Figure 16-17, with the lid replaced.

Conclusions

I'm betting that you had some trouble trying to control the solvent in this project. The applicator is difficult to use, as it allows liquid to escape unpredictably. The only way to deal with this problem is by practicing with the applicator.

Here are some take-home messages about using solvent with plastic:

When you are working with transparent plastic, you can see immediately if you have made a good joint. Opaque plastic leaves you wondering.

Rough-sawn or uneven edges do not work well with solvent, because it can't fill gaps. Edges must fit well and must be smooth.

If a solvent joint is good, it can be quite strong, as the plastic merges together.

In a conventional glued joint, temperature changes can cause stresses from unequal contraction and expansion of the glue and the surfaces that it has joined. In a joint created with solvent, this should not occur, as the joint consists of pure plastic after the solvent has evaporated.

Personally, I like the look of glued plastic, but achieving the result is difficult, and I never quite trust the joints. Is there an alternative? Indeed there is, as I will show in you in Chapter 17.

Figure 16-15. Ready to apply cement to the base of the box from the inside.

Figure 16-16. The finished box with lid removed.

Figure 16-17. The finished box with lid replaced.

Chapter 17
Another Better Box

NEW TOPICS IN THIS CHAPTER

- ABS plastic with sheet-metal screws

YOU WILL NEED

- Two-by-four pine, any condition, length 12"
- One-by-six pine, a few knots, not warped, length 12"
- Plastic sheet, white ABS, ⅛" thick, 12" wide, length 12"
- Sheet-metal screws, stainless steel, #2 x ½", flat head, Phillips, quantity 13
- Screwdrivers, miniature set including Phillips size 1

Also, as listed previously: Tenon saw, miter box, trigger clamps, ruler, speed square, rubber sanding block, work gloves, awl, utility knife, deburring tool (optional), electric drill and drill bits, countersink, safety glasses (optional), plywood work surface, and sandpaper.

Check the Buying Guide on page 248 for information about buying these items.

Just in case you found the gluing process difficult in Chapter 16, I'm going to show you how to use screws with ⅛" ABS in a way that takes advantage of its uniform texture. Wood grain always tends to push a drill bit or a screw in one direction or another, even when you are working with high-quality hardwood. Because plastic has no grain, it doesn't create that kind of problem, and you can achieve greater precision, even with a handheld drill.

The screws that I want to use with ABS are tiny, because I plan to insert them into the edge of the ⅛" sheet. Yes, this can be done! But #2 screws are uncommon, and you will probably have to order them online. I suggest stainless steel, because in my experience they have sharper threads and are better made. They are not expensive.

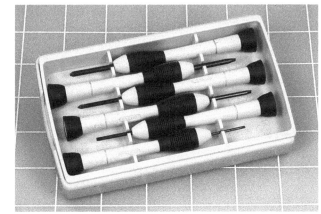

Figure 17-1. A set of miniature screwdrivers including the Phillips size 1 that you need for this project.

Assuming your screws have Phillips heads, you need a miniature size 1 Phillips screwdriver. A hardware store may not stock this as a separate item, but it should be included in a set such as the one shown in Figure 17-1.

As you might imagine, #2 screws are thin—just under ³⁄₃₂" in diameter, including the threads. But a ⅛" piece of plastic is thicker than that. How much thicker? Let's figure it out. ⅛" is the same as ⁴⁄₃₂", so if you could insert a ³⁄₃₂" screw in the edge of a piece of ⅛", the screw would allow ¹⁄₆₄" of plastic to remain either side of it. Not a whole lot, but, it can work.

What about a pilot hole? In my experience a #2 screw grips quite well in a ⁵⁄₆₄" hole when you are using it with ABS. Can you drill that hole in the edge of a ⅛" sheet? Figure 17-2 suggests that it should be possible.

64ths of an inch

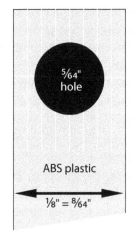

ABS plastic

⅛" = ⁸⁄₆₄"

Figure 17-2. How a pilot hole can be drilled into the edge of ⅛" plastic.

So much for the theory. If this sounds too ambitious, bear in mind you can scale up the whole project. You can use ABS that is ³⁄₁₆" thick instead of ⅛", and multiply all the dimensions by 1.5. The screws that hold it together should be 1.5 times as long, but they don't need to be 1.5 times thicker. Or you can go all the way to ABS that is ¼" thick.

On the other hand, bear in mind, I am not a highly skilled crafts person. In fact, for my whole life, I have had a slight tremor in my hands. I tend to think that if I can build this, you can probably build it too.

To find out, let's try a feasibility study.

An ABS Test

You should have some small pieces of scrap ABS left over from the previous box. One of them can be clamped vertically, in an arrangement such as that in Figure 17-3. ABS is so soft in comparison to a steel bit, you don't need to mark the edge of the ABS with an awl. The bit will start the hole easily, and you can tell if it is centered on the edge of the plastic just by looking at it.

Put a ⁵⁄₆₄" bit in your drill, and run the drill as slowly as possible, so that the bit is barely turning. Touch the tip of it where you want the hole to be, and adjust it a little each way till it's centered. The bit should make an indent such as the one shown in Figure 17-4.

If the indent isn't centered, abandon that location and try again further down the edge. But if the indent does seem to be centered, you're ready to make the actual hole. Check that your drill bit is exactly parallel with the vertical sides of the ABS, and press a little harder, while still running the drill very, very slowly. Keep your hands out of the way of the drill bit, in case it breaks through the side of the plastic.

Because ABS has no grain, there's nothing to push the drill off its track. The hole will make itself, and it only needs to be ½" deep. Little threads of ABS will come out, as in Figure 17-5. Remember that this is a ⁵⁄₆₄" drill; the photograph was taken with an extreme closeup lens.

The finished hole should have well-defined edges, as in Figure 17-6.

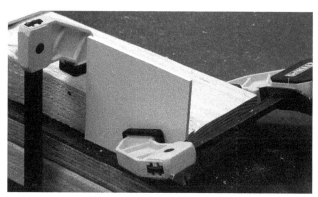

Figure 17-3. Clamping a piece of scrap ABS for a feasibility test.

Figure 17-4. Starting a hole in the edge of the sheet.

Figure 17-5. Drilling in progress.

Figure 17-6. The finished hole.

Take another scrap piece of ABS and mark two lines ¹⁄₁₆" apart on its smooth side, parallel with an edge and ¹⁄₁₆" away from it, as in Figure 17-7. A pen line may be too thick to do this accurately, so I used a pencil. Because it's more difficult to judge the position of a drill bit on a wider surface, I made a mark with an awl.

You need to make a screw hole on the center line. Use the same ⁵⁄₆₄" bit, because the next size up will be too big. Figure 17-8 shows the hole being drilled, and Figure 17-9 shows the hole finished.

Figure 17-7. Two parallel pencil lines, separated by ¹⁄₁₆".

Figure 17-8. Making the screw hole.

Figure 17-9. The screw hole complete.

Figure 17-10. The hole after countersinking.

Now you need to countersink the hole, as you'll be using a flat-headed screw. Turn the sheet over, as the textured side is usually the outside of any object that you build with ABS. The countersunk hole is shown in Figure 17-10.

You'll notice that the beveled edge has broken through the edge of the plastic. I'll deal with this issue in a moment.

Using your miniature screwdriver, drive a #2 screw into the hole that you just made. It's not a wood screw, so its thread will pull it all the way in. When its head is flush with the surface of the ABS, keep turning the screwdriver, and the threads at the top of the screw will round out the inside of the hole. The screw starts turning freely—just as if it had a smooth shank.

Figure 17-11. A Philips #1 screwdriver is necessary to fit the #2 screw.

Insert the tip of the screw in the first hole that you drilled, and the threads will bite just enough to pull the screw in, without breaking through the outer surface of the plastic or even making it swell visibly. But if you over-tighten the screw, it will strip the threads. A gentle touch is needed. See Figure 17-11.

The pieces of plastic joined with a screw are shown in Figure 17-12. But, you see the problem. The screw head is a little too wide, and overlaps the edge of the plastic. Does it matter? That depends. When I'm making something small that is supposed to look very precise, I like everything to be just right. In any case, I'm not sure screws look good so close to the edge. So I designed a jewelry box with the sides recessed, as shown in Figure 17-13. I think the screws look nicer that way.

Figure 17-12. The screw head doesn't quite fit inside the edge.

Box Design

The box is shown with its lid open in Figure 17-14, and with some of the parts separated in Figure 17-15.

The other projects that I have described in this book didn't attempt to be stylish, because everything has to be simple when you're learning the basics. In this one I tried to make it a little more visually interesting. Also, by extending the base outward and standing the sides of the box on it, I was able to hide four more screws, which are under the base, going up into the side pieces.

Inevitably, when something looks more visually interesting, it requires a little more work. This box requires nine pieces of plastic, to be cut out of two strips of differing widths. On the other hand, you won't have to struggle with solvent cement.

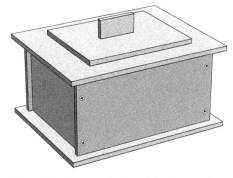

Figure 17-13. A rendering of the jewelry box.

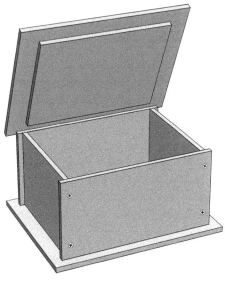

Figure 17-14. The box with its lid open.

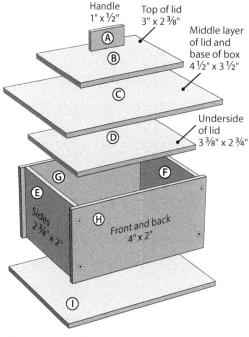

Handle
1" x ½"

Top of lid
3" x 2⅜"

Middle layer of lid and base of box
4½" x 3½"

Underside of lid
3⅜" x 2¾"

Sides
2¾" x 2"

Front and back
4" x 2"

Figure 17-15. The parts separated and identified.

The pieces that you need to cut are shown in Figure 17-16, with letters that refer back to Figure 17-15.

Figure 17-16. Pieces of plastic to cut and drill.

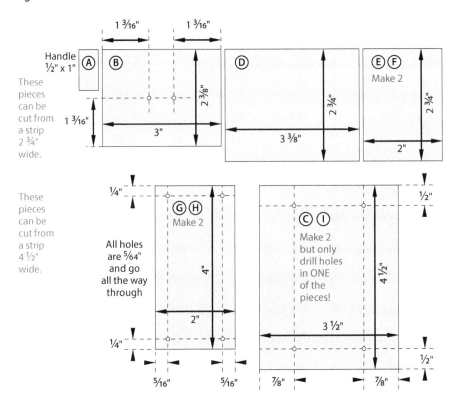

To make pieces in the top row, cut a strip 2¾" wide in the same way as in Chapter 16, and then cut the width of each piece out of it in your miter box (remembering to make two copies of the one labeled E and F). Then trim pieces A and B to the right height.

For pieces in the lower row, use a strip of plastic 4½" wide, then cut the width of each piece out of it (remembering to make two copies of each), and trim the pieces labeled G and H to 4" high. My miter box is just wide enough for a 4½" strip; I hope yours is, too.

Next drill the holes in pieces B, G, and H. I suggest you mark the locations on the smooth side, and drill through to the textured side. Making marks, and cleaning them later, will be easier on the smooth side. Because all the pieces are symmetrical, you can turn them over later if this is required by the assembly process.

Countersink each hole on the textured side. *Don't drill piece C yet.* Just mark where the holes are supposed to be. The locations may have to be adjusted slightly.

Begin assembly with piece H, the front of the box and a side of the box, such as E. In Figure 17-17, piece E has been clamped vertically to the side of a two-by-four, which is clamped to the work area. Align piece H with piece E, and with a sharp pencil, mark through the screw holes to establish locations for pilot holes.

Set the longer piece aside, and drill ⁵⁄₆₄" pilot holes where you see the pencil marks, as in Figure 17-18.

After the screws have been inserted, the two pieces are shown joined together in Figure 17-19.

Use the same procedure to add more pieces, and you'll find that the sides of the box are just far enough apart to straddle a two-by-four, as in Figure 17-20.

Figure 17-17. Marking the locations for pilot holes.

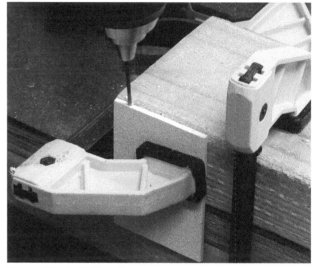

Figure 17-18. Drilling pilot holes.

Figure 17-19. The first two pieces, joined.

Figure 17-20. Continuing to assemble the box.

When you have joined all four sides, stand them on the base, which should be smooth-side up. Measure the margins around the sides of the box, to check that they are centered. Look to see whether the locations for the screws through the

Figure 17-21. Pilot holes to match screw holes in the base.

Figure 17-22. Marking the remaining two pilot holes while the base is temporarily held in place.

base will match the positions of the sides. Draw around the inside and outside of the box with a pencil, if this will help to clarify the situation. Relocate the hole positions slightly, if necessary, and then make your holes through the base.

Turn the box over, and rest the base on it. Make pencil marks through the holes that you drilled, then remove the base and drill pilot holes, with the box clamped as in Figure 17-21.

Install screws temporarily in the first two holes, as shown in Figure 17-22, to stabilize the assembly while you mark the remaining two holes.

You may find that some of the screw holes don't line up precisely with your pilot holes, although this is less likely to happen with ABS than it was with plywood, because ABS doesn't contain grain that pushes your drill off track. Fortunately ABS is so soft and flexible, if a screw doesn't quite line up, the screw hole will widen slightly to allow the screw to go in anyway.

You may remember that in Chapter 10, you measured all the positions of pilot holes, and drilled them before assembly. That was a learning experience, demonstrating the difficulty of doing precise work, especially with plywood. In this project, marking through a screw hole to the position of a pilot hole should make the process of assembly easier.

The last two screws will attach the handle to the top surface of the lid. Drilling pilot holes in the handle requires careful clamping, as in Figure 17-23.

But, it can be done. See Figure 17-24.

Check the bottom layer of the lid (piece D) to see if it will fit within the sides of the box, and sand it down as much as necessary, as you did with the glued box in Chapter 16. When it fits easily, you need to assemble the three layers of the lid. This isn't possible with screws, because the layers of the lid aren't thick enough for you to sink screws into them. You need to use solvent cement, which is easy on such a wide, flat area. I placed masking tape to indicate where each piece should be glued, as in Figure 17-25.

When dribbling cement onto a flat surface, be careful to place the surface horizontally. Otherwise the cement will run off, with unexpected consequences. Remember to use eye protection and hand protection during the gluing process, and work in a well-ventilated area.

Figure 17-23. Drilling pilot holes in the handle.

Figure 17-24. Completed pilot holes in the handle.

Figure 17-25. Masking tape shows where to place the top of the lid when it is glued to the middle layer of the lid.

The finished jewelry box is shown in Figure 16-26.

I tend to think that something small can look more attractive than something big, even though it may be built in the same way. Screws and butt joints look okay to me on the scale of the jewelry box, especially because the edges of ABS don't have the rough, splintery look of plywood.

I have to admit, though, that fabricating larger objects is easier. The main reason I specified ⅛" ABS for this project is that I like to use the same materials in multiple projects, and ⅛" was quick and easy to bend in Chapter 15.

So far, I have only used ABS, but of course there are many other plastics that you can use, especially if you want transparency. The next chapter shows you some examples.

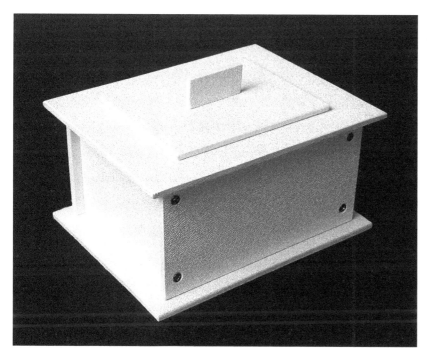

Figure 17-26. The finished jewelry box, with four visible screws and four hidden screws. Do the screws spoil its appearance?

Chapter 18

Transparency

NEW TOPICS IN THIS CHAPTER

▪ Polycarbonate and acrylics

See next page for a materials list.

The last three chapters were devoted primarily to ABS, because I think it's the easiest plastic to work with, and it's less expensive than some alternatives. It has one big drawback, though: transparent ABS is relatively rare.

Transparent Options

Two popular types of transparent plastic are **polycarbonate** (often known by the brand name **Lexan**), and acrylic (sold under brand names such as **Lucite**, **Perspex**, and **Plexiglas**).

Some advantages of acrylic plastic:
▪ Less easily scratched than polycarbonate.
▪ More resistant to ultraviolet (polycarbonate gradually yellows).
▪ Excellent clarity, and can be restored by polishing.
▪ Cheaper than polycarbonate.
▪ Available in numerous shapes and colors.

Some advantages of polycarbonate:
▪ Much greater impact resistance—less brittle than acrylic.
▪ Doesn't tend to chip while you are working on it.
▪ Easier to bend than acrylic.

Among the many applications, acrylic plastic is used in helmet visors, helicopter windows, store displays, signage, and underwater windows. Polycarbonate is used for CDs and DVDs, lenses, safety glasses, water bottles, and instrument panels.

Polycarbonate in a Picture Frame

You can buy untinted polycarbonate from a hardware store, where precut pieces are sold as a substitute for window glass. You'll find that both sides of each piece are covered with adhesive protective film, which you leave in place while you saw or drill the plastic. After these operations, you peel the film away. You also have to remove the film if you're going to use heat to bend the plastic.

Perhaps you remember the pentagon-shaped frame that I featured in Chapter 5. Back then, I suggested that a piece of polycarbonate could be installed as a substitute for glass. It's more expensive, but much easier to use and safer to handle. You can just cut it with a saw.

The protective film on polycarbonate is usually transparent, allowing you to look through it while making measurements. I used a fine-point permanent marker to trace the inner boundary of the frame, as in Figure 18-1.

After drawing the outline, you can saw the polycarbonate in exactly the same way as you sawed ABS. See Figure 18-2.

Removing the protective film is easy, as shown in Figure 18-3.

After sanding its edges, I placed the transparent plastic in the back side of the frame, behind the wooden trim that was glued there previously. Peeling off the protective film tends to charge the plastic with static electricity, which attracts workshop dust. Wiping the polycarbonate with a tiny amount of dishwashing liquid helps to remove the charge, leaving a nice clean, reflective surface, as in Figure 18-4.

Figure 18-1. Drawing on the plastic film that protects a sheet of polycarbonate, using a fine-point permanent marker.

Figure 18-2. Ready to saw the polycarbonate.

Figure 18-3. Removing the protective film.

Figure 18-4. The plastic panel, cleaned and installed.

YOU WILL NEED

- Two-by-four pine, any condition, length 12"
- One-by-six pine, a few knots, not warped, length 12"
- Plastic sheet, white ABS, ⅛" thick, 12" wide, length 6"
- Plastic sheet, clear polycarbonate, 1⁄16" or 3⁄32" thick, 12" wide, length 24"
- Plastic rod, round, PVC, 3⁄16", length 360" (optional)
- Plastic caps for ½" plumbing, PVC, quantity 60 (optional)
- Sheet-metal screws, stainless steel, #2 x ½", flat head, Phillips, quantity 4

Also, as listed previously: Tenon saw, miter box, trigger clamps, ruler, fine-point permanent marker, speed square, rubber sanding block, work gloves, awl, utility knife, deburring tool (optional), electric drill and drill bits, countersink, miniature screwdrivers, heat gun, heat-resistant gloves (optional), ceramic tiles, solvent cement and applicator, chemical-resistant goggles, chemical-resistant gloves, clean rags, cardboard 12" x 12", plywood work surface, and sandpaper.

Check the Buying Guide on page 248 for information about buying these items.

In case you may have felt that the cartoon that I used previously was disrespectful to people who have pointy heads, I found a picture that may be more appropriate. See Figure 18-5.

A piece of corrugated cardboard is sufficient to go in the back of a frame. I secured it with ½" brads around the edges. That completed this leftover project.

Mixed Media

The solvent that you used in Chapter 16 dissolves polycarbonate as well as ABS. Does that mean you can cement the two types of plastic together? Absolutely!

Figure 18-5. An appropriate image from Google Earth.

When someone builds a little electronic circuit, the finished result can be installed in a **project box**. This is just a little box that can contain the electronics, often along with a 9-volt battery, while switches, buttons, and LEDs can be mounted on the top. Yes, this is another box project—but useful, and a good way to demonstrate that screws, glue, and bending can be combined to simplify a design, while mixing different plastic media.

Project boxes tend to look pretty dull, but they don't have to. Why not build one that is partially transparent, to reveal the components inside? Figure 18-6 shows what I have in mind. No doubt it can be used for other purposes, too, such as displaying collectibles.

This only takes about an hour to build. You need a piece of polycarbonate 12" x 4", a piece of ABS 12" x 4", four screws, and some solvent cement with an applicator, which I introduced in Chapter 16.

Figure 18-6. A box with transparent front, back, and top.

Begin with the ABS. From the strip 4" wide, cut two copies of the shape shown in Figure 18-7. Because these shapes will be on opposite sides of the box, you should flip the pattern left-to-right for the second piece, so that the textured side of each piece can face outward.

Drill the holes and use a countersink on the outside (textured) surface, to prepare the holes for #2 flat-headed screws. Round the two top corners of the ABS by sanding them till you like the look of them.

Now take the piece of polycarbonate and insert it between your heat-protective tiles, leaving just over 4" sticking out. Apply your heat gun to an exposed strip that is slightly wider than ½" because this has to be a gentle curve rather than a sharp bend. Polycarbonate has a higher melting temperature than ABS, but you are using a piece only ¹⁄₁₆" or ³⁄₃₂" thick, so it should lose its stiffness after about a minute.

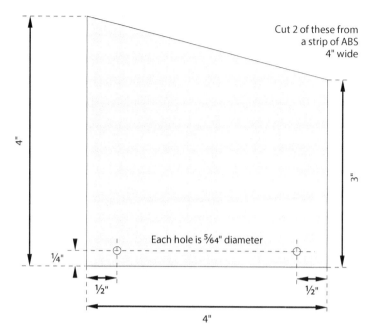

Cut 2 of these from a strip of ABS 4" wide

4"

3"

Each hole is ⁵⁄₆₄" diameter

¼"

½"   ½"

4"

Figure 18-7. A plan for the two ABS sides of the box.

When the polycarbonate is quite limp in the hot area, remove it from the tiles and wrap it around one of the ABS pieces that you prepared, as in Figure 18-8. Make sure the bottom corner of the polycarbonate lines up with the bottom corner of the ABS. Pull the polycarbonate tight around the ABS and wait till it sets. This will probably require both of your hands, but if someone is with you, they can spray the plastic with a little water to expedite the cooling process.

Put the polycarbonate back between the tiles and apply heat about 4⅛" further along the strip. Then wrap it around the next corner in the same piece of ABS.

Figure 18-8. Wrapping the first bend in the polycarbonate around the piece of ABS.

The polycarbonate will extend beyond the bottom edge of the ABS, so you'll need to mark it and saw off the excess after the bend becomes rigid. The finished polycarbonate should look something like the example in Figure 18-9.

Figure 18-9. The polycarbonate has been bent and trimmed.

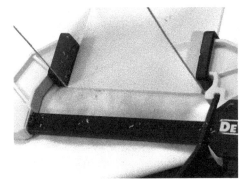

Figure 18-10. Add solvent around the three edges of the ABS section to bond it with the polycarbonate.

Figure 18-11. One side has been cemented. The polycarbonate has not been cleaned, yet, so it still shows specks of plastic dust.

Clamp the polycarbonate around the ABS piece that you have been using, and place it on some clean rags, as in Figure 18-10. You could use screws to assemble these pieces, but I think solvent is a better-looking option, and quicker.

Check Chapter 16 for the procedure and cautions regarding solvent. Please don't ignore the need to wear gloves and eye protection.

Trickle solvent in around the clamped edges, and surface tension will draw it into the crack. Pick up the assembly in case some solvent has run through and into the rags. Within a minute or two, the joint should be strong enough to be unclamped, as in Figure 18-11.

You should be able to cement the second side in the same way, if your bends in the polycarbonate have been symmetrical.

Reheat the polycarbonate if it really won't fit.

After you have cemented the second side, it's time to cut one more piece for the base of the box. This will be recessed inside the other pieces. One dimension of the base will be 4", because your pieces of ABS are 4" wide. You can't be sure of the other dimension, because of small variations that tend to occur during the bending and gluing process. Therefore you should measure the actual distance between the two sides, then cut the base to fit. Make it a fraction bigger than necessary, and sand it till it slips in without forcing.

You will use screws to hold the base in position, because the box won't be much use if you can't get inside it. Turn the box on its side, fit the base in, and nudge it along till you see the edge of it through the holes in the side facing you. Use a sharp pencil to mark the edge of the base through the holes, in exactly the same way as when you inserted the sides of the box in Chapter 17 (assuming you built that project).

I guess I have to admit that when I was building my version of this box, I forgot to drill holes in the side pieces when I cut them. You can see the absence of holes in the previous photographs. I could have rebuilt the project and photographed it again, but I decided that it's useful to demonstrate that everyone makes mistakes. I hope your memory is better than mine, because it's easier to drill the holes before you start to assemble the pieces than after they have been glued in place.

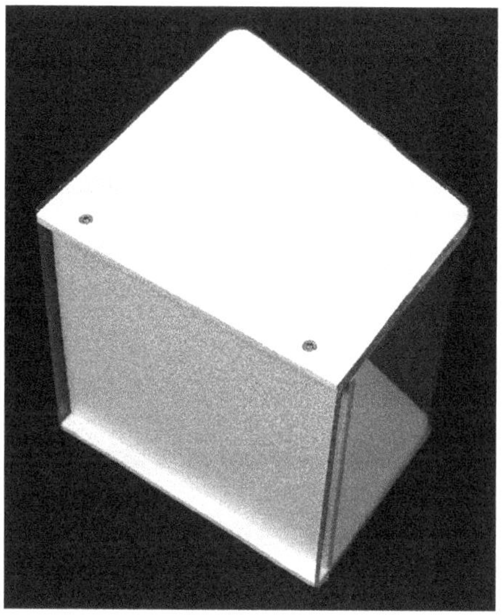

When you have marked the edge of the base through the screw holes, remove it, clamp it in a vertical position, and drill pilot holes into the edge, as was described in Chapter 17. Then put it back in the box and insert the screws, as in Figure 18-12.

Figure 18-12. One side of the box has been attached to the base with two screws.

The last step is to turn the box over and mark the opposite edge of the base through the holes on the other side. Remove the first two screws, drop the base out of the box, and once again, drill pilot holes on your pencil marks.

Before you reinsert the base on a permanent basis, you'll need to clean up your work area and then wipe both sides of the polycarbonate to make it dust-free. The result should look like Figure 18-6.

The dimensions can be changed easily if you want to use the same basic design to build a different size of box. Also, if you have trouble inserting the screws into the edge of ⅛" ABS, you can upgrade to a sheet ³⁄₁₆" or ¼" thick, without changing any of the steps in the fabrication process that I just described. Only the width of the base will be affected, and you'll be cutting that to fit anyway.

More Shapes

For all of these projects, you've been using plastic in sheets. Suppliers also sell various types of plastic in rods, bars, and tubes, for applications ranging from laboratory equipment to store displays.

You can also obtain molded shapes such as discs, squares, cubes, and spheres, especially in the world of acrylic plastic. Some acrylic discs and spheres measuring ⅝" and ⅞" diameter are shown in Figure 18-13. You can find them from many sources on eBay, and may also see

Figure 18-13. A gathering of acrylic spheres.

them in some crafts stores. Some people use small pieces of molded acrylic, in a rich variety of colors, to make jewelry. You should find quite a few jewelry crafts books if you check online or at your local bookstore.

More Plastics and Applications

Many other types of plastic exist, such as the PVC (polyvinyl chloride) that is often used in water pipes and electrical conduits. You can also get PVC rods just ³⁄₁₆" in diameter that are used in 3D printers.

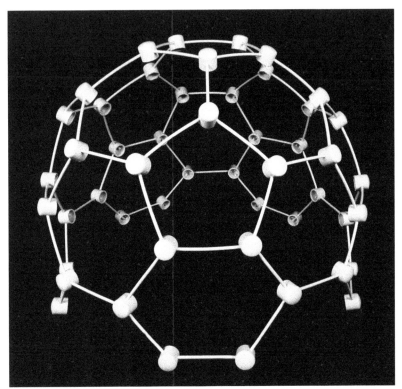

Figure 18-14. An all-PVC geodesic framework.

Figure 18-15. A geodesic soccer ball.

I used some of these rods to make the structure in Figure 18-14. The rods are joined in PVC caps intended for ½" plumbing, cheaply available in bulk. I drilled three holes in each cap, inserted the rods in the holes, and applied PVC plumbing cement, wiggling the rods to get the cement into each joint. This is a solvent, but easier to apply than the type you used with ABS, because it is thicker and takes longer to set. Most PVC cements are brightly colored, but I found one that is white when I searched online.

Does this shape look familiar? It's identical with the geometry of a soccer ball, as shown in Figure 18-15. This came to be known as the Buckminster ball, although Buckminster Fuller did not actually design it. It was introduced for the 1970 soccer World Cup, and was used for several decades.

The rest of my suggestions, below, are just thought experiments. The materials are not included in the buying guide.

Can you think of applications for this kind of plastic structure? Maybe you could wrap it in chickenwire to make a bird cage—although for a parrot, you should substitute a metal frame to prevent the bird from eating its way out.

If you scaled it up, you could use lengths of 1½" PVC pipe and 3" caps to connect them, and you'd be able to build a little greenhouse—although the frame would be stronger if it was based on triangles instead of pentagons and hexagons. Triangles were the basis of the true geodesic-dome geometry that Fuller established many decades ago. You can find a lot of web sites with precise dimensions for dome structures.

PVC is also available in thin sheets, and the cement that works with them will also work with thin vinyl, which you can buy by the yard from fabric stores. This suggests to me that you could make a geodesic a lamp shade using panels of transparent PVC joined with gray vinyl, as in the rendering in Figure 18-16. In this design, gray diamonds would be glued over the edges of white pentagons, to hold everything together and create the star pattern.

Alternatively, figure 18-17 suggests mitering 12 pentagon-shaped wooden panels, each with a circular hole that could be filled with translucent plastic. The mitering is not quite as challenging as it seems, because you can look up the miter angle in a standard geometry textbook. It is approximately 58¼ degrees on every edge. Maybe you could design a jig to simplify the sawing.

The basic shape in figures 18-16 and 18-17 is called a *dodecahedron*, which is one of the five *Platonic solids*, named after the Greek philosopher Plato. The full set is shown in Figure 18-18. What other concepts do these shapes suggest?

Plastic, with or without wood, opens up all kinds of possibilities. I've suggested some, but there are many more. In particular, one topic that I haven't dealt with is color. I'll get to that in the next chapter.

Figure 18-16. Technically, this design is built on top of a dodecahedron, which you may recognize as a 12-sided die in some role-playing games.

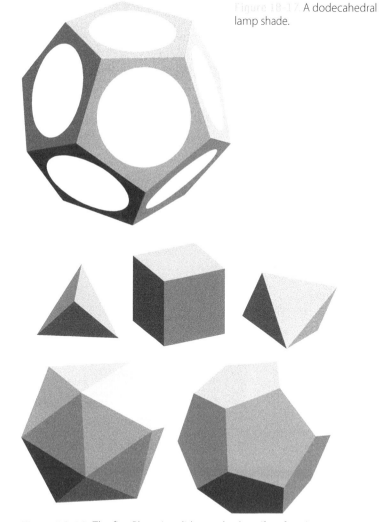

Figure 18-17. A dodecahedral lamp shade.

Figure 18-18. The five Platonic solids: tetrahedron (four faces), hexahedron (six faces), octahedron (eight faces), icosahedron (20 faces), and dodecahedron (12 faces).

Continued on the following page

Natural wood can be beautiful, so long as you enjoy shades of brown. If you want a more vivid palette, paint is always an option—but plastic can be much easier. Opaque colored plastics have a rich, uniform luster, while transparent and translucent plastics have unique attributes when they are backlit. The two projects in this chapter show what I mean.

Colored Acrylic

For a really wide range of colors, acrylics are probably the best choice. I put together a few samples in Figure 19-1 and placed a bright white surface behind them. If I had put a lamp behind them, you would see the shape of the lamp very clearly, because although these colors are tinted, they are also transparent.

Figure 19-1. Six colors of transparent acrylic, with other colors formed where they overlap.

This raises the need to distinguish between **opaque**, **translucent**, and **transparent** materials. You may run into some confusion about these terms, even on web sites that sell plastics. I once spent 15 minutes on the phone with a supplier, pointing out that some acrylics on their site were listed as "translucent" while a video on the same site described exactly the same products as "opaque." So which was correct?

"Well, the light just doesn't get through," the sales person told me.

"So it's opaque, then."

"No, it's translucent."

But if it's translucent, some light does get through, because "trans" means through and "lucent" means glowing. Clearly (no pun intended), we need some definitions.

Transparent plastic may be **clear** or **tinted**. If it is tinted, you can still see objects behind it. Sunglasses are transparent in this way. Transparent tinted plastic is suitable to be placed in front of a floodlight, because it only changes the color of the light, without diffusing the beam. The intensity of the light is reduced, but only because some wavelengths are being blocked.

Translucent plastic allows light to penetrate it, but not as much, and the light is *diffused*. Translucent acrylic may be used for signage that is illuminated at night by fluorescent tubes behind it. In an ideal world, a supplier would state what percentage of light is allowed through, and indeed some suppliers do give a number for their white translucent sheets—but not for the colors. However, we do know that the thicker the sheet is, the more it will tend to block the light. Therefore, you should use sheets that are no more than ⅛" thick when you want as much light to be transmitted as possible.

Opaque plastic transmits virtually no light at all, regardless of its color. In fact the color of the plastic is only visible if there is a light source reflecting from the front it. Any source behind the plastic will be blocked. The white ABS that you have been working with is opaque.

An Alphabetical Nightlight

To dramatize the distinction, I'll describe a project that mixes transparency and translucency. The purpose of this project is to build a night light.

Suppose someone named Erica has a birthday coming up, and you suspect she might enjoy a hand-made night light in which her initial becomes visible when the light is switched on. How can you arrange for a letter E to appear in this magical way?

In Figure 19-2, you see the enclosure made from white (opaque) ABS, with a grayish piece of polycarbonate at the front end. That's all you can see, because the photograph was taken with an external light source.

Wood screws, #8 x 1½", flat head, Phillips, quantity 2

Deck screws, 2¼", quantity 5

Also, as listed previously: Tenon saw, miter box, trigger clamps, ruler, speed square, rubber sanding block, work gloves, awl, utility knife, metal files, deburring tool (optional), electric drill and drill bits, countersink, screwdriver or electric screwdriver, heat gun, heat-resistant gloves (optional), ceramic tiles, solvent cement and applicator, chemical-resistant goggles, chemical-resistant gloves, clean rags, cardboard 12" x 12", plywood work surface, and sandpaper.

Check the Buying Guide on page 248 for information about buying these items.

Figure 19-2. An opaque ABS enclosure with a translucent front panel and something behind it that isn't easy to see, yet.

In Figure 19-3, the external light has been dimmed, and a light inside the enclosure has been switched on. Not only is the E revealed, but you can see three distinct colors. How is it done?

The front panel is actually a piece of clear polycarbonate that was sanded on both sides with 220 grit sandpaper. The sanding should be done with circular motions so that it looks the same when light falls from different directions. The thousands of tiny grooves cut by the grit diffuse the light so that the plastic becomes translucent instead of transparent. You can see this in Figure 19-4.

Figure 19-3. The letter E, revealed.

Figure 19-4. A piece of clear polycarbonate that has been sanded to make it translucent.

Why not just use off-the-shelf white translucent acrylic? Because it turns out to be a little too dense, and it diffuses the light too much, making everything blurry. Sometimes we have to do things ourselves to get exactly the effect that we want.

Overlapping Colors

To create the colors behind the translucent panel, you can use a coping saw to cut shapes out of blue transparent acrylic and magenta transparent acrylic, as shown in Figure 19-5.

Figure 19-5. Two shapes to be cut out of two colors of transparent plastic.

Because these pieces of colored plastic are transparent, light passes through them cleanly, so that the edges of the E are sharply defined. Also, the pieces create a third color when they overlap. This concept is illustrated in Figure 19-6, where the two shapes are shown partially overlapping. The diagram is just a simulation, but is a good imitation of what actually happens. Each individual color subtracts some wavelengths from the visible spectrum, and where light passes through both of the colors, almost all the wavelengths are blocked, so that the combination is a very dark blue.

When the two layers are aligned, the result is shown in Figure 19-7.

Plans to cut the magenta plastic are shown in Figure 19-8, and to cut the blue plastic in Figure 19-9. But wait—you may not happen to have two pieces of acrylic ready-cut into pieces 4" x 4" square, and I haven't explained how larger pieces of acrylic are often chopped into smaller pieces.

Figure 19-6. The two shapes partially overlapping.

Figure 19-7. The layers aligned.

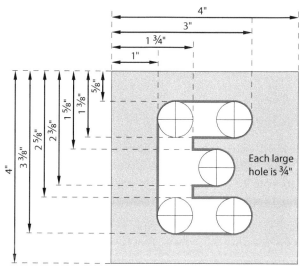

Figure 19-8. Plans for the magenta acrylic square. Red lines indicate saw cuts after the holes have been drilled.

Each large hole is ¾"

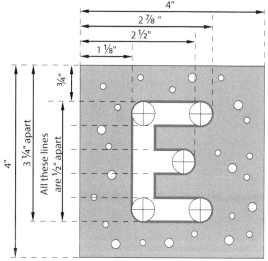

Figure 19-9. Plans for the blue acrylic square.

Each large hole is ½". Small holes are ⅛" and ³⁄₁₆" located randomly.

Snapping Acrylic

Because acrylic is hard and brittle, people often prefer to break it along a line instead of sawing it. A special knife of the type shown in Figure 19-10 is available from plastic suppliers.

Figure 19-10. A knife for scoring sheets of acrylic plastic.

When you drag the knife across the plastic toward you (alongside a steel ruler, to provide guidance), its reversed blade digs in and scores a line. Make a few strokes, carefully following exactly the same line each time, and be sure that the line extends all the way to each edge. Figure 19-11 shows a score mark in a scrap piece of acrylic. I put it on white paper to make it more easily visible.

Figure 19-12 shows the plastic positioned ready for snapping, while Figure 19-13 shows the plastic successfully snapped. Because the edge can be sharp, it's a good idea to sand it or debur it right away.

Figure 19-11. A piece of acrylic scored by a plastic-cutting knife.

Figure 19-12. To snap the acrylic, push down on the part that sticks out.

Figure 19-13. Immediately after snapping.

Time for a Coping Saw

I'll assume that you now have two 4" x 4" squares for the night-light project. If you look at my design for the letter E, I deliberately rounded the corners and the ends of the letter, to make it easier to fabricate. All you have to do is drill five holes in each piece, score lines between them with an awl, and then saw along the lines. See Figure 19-14.

You may need to remove the protective paper from the piece of acrylic, so that you can get a clearer view of what you are doing. This requires careful handling of the plastic, and to avoid scratching it, you can clamp it inside two layers of soft rags.

Ideally, you need a drill bit that can cut clean ½" and ¾" holes. The best option is a Forstner bit. A spade bit will be cheaper, but may make a mess or break the plastic, with hazardous results. Always bear in mind that acrylic is brittle and chips easily.

Instead of drilling holes, you can use the coping saw to cut around all the curves, but that will entail some extra work, and you'll have trouble making the tighter curves precisely. Take advantage of the ability to rotate the blade of the coping saw in its frame, by loosening and retightening the handle. This will allow you to cut in more than one direction without unclamping and turning the plastic.

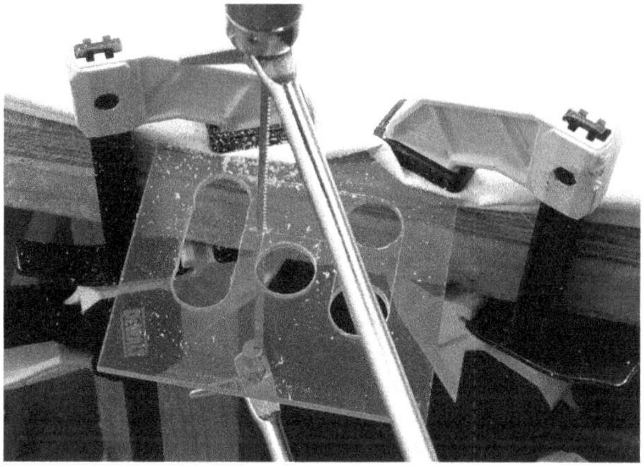

Figure 19-14. Cutting along lines that were scored between holes defining the letter E.

To smooth the edges after making all the cuts, I suggest a semicircular metal file, which is well suited to the hardness of acrylic. If you used the coping saw instead of a drill bit to make the ½" and ¾" holes, you'll probably need a circular file to smooth those saw cuts. Files are useful because a deburring tool tends to chip acrylic, while sandpaper risks scratching the surface.

The Enclosure

The three layers are shown in Figure 19-15, and as you can see, their shapes are visible when they are close to the sanded polycarbonate. A translucent panel always reveals the outline of a shape more clearly when the two pieces are in contact. As the shape moves further away, it becomes blurred and indistinct. The shapes are also visible in this photograph because white paper is reflecting light up through them.

How to install these layers in an enclosure? You can make bends in a strip of ABS by heating it and curving it around the front panel. Then clamp and cement the ABS to the panel, in exactly the same way that you cemented the top, front, and back of the box in Chapter 18.

Figure 19-15. The three layers for the night light.

To install the colored layers, you may be tempted to cement them, too. But runs and splashes of solvent are difficult to control, so you can just place the squares in the enclosure behind the front panel of polycarbonate, and cement a couple of small pieces of ABS behind them, to stop them from falling out.

There's one challenging aspect of this project that I haven't mentioned. Letter E is relatively easy to create, because doesn't have any strokes that form a closed loop. In the case of letters A, B, D, O, P, Q, or R, when you cut their strokes, the centers will fall out. To deal with this problem, you could use solvent to hold the center of a letter against an additional square of clear plastic. Or, you can just make a night light for someone who has an easier initial!

Illumination

If you build this project, you have to decide how to illuminate it. A 15-watt appliance bulb is likely to be too bright for a night light. You can buy a *candelabra bulb* rated for five watts. It will probably have a small E12 base. You can screw this into an E26 to E12 converter, to obtain a regular-size US screw base. Then insert that into a light-socket-to-plug adapter, which is very cheaply available, and plug that into an extension cord.

A five-watt bulb generates very little heat, but still, you should leave the back end of the enclosure open to provide some ventilation. Be sure to protect the ABS from coming into contact with the bulb.

Even a five-watt bulb may be too bright and too warm. Really, a better choice would be an LED that runs from low voltage and generates negligible heat. You can buy individual 5mm white LEDs that use less than 20 milliamps—that's 0.02 of an amp, enabling the night light to be battery powered. Two or three C cells, wired in series, should be able to run an LED for more than a month at eight hours per night. So, this little project could be portable and self-contained. You'd just need to add a switch.

I'm skipping over some details. For instance, an LED has polarity, meaning that electric current has to run through it in the right direction. The LED will also need the right voltage, within fairly strict limits, and a series resistor. You'll have to shop for these items online, with a risk of buying something that doesn't quite fit the specification, and won't work properly.

These issues can be dealt with, but I don't have space to explain them here. If you want to build any project that uses electronics, I think you should pick up an introductory guide on that subject. I can't resist mentioning my own book, *Make: Electronics*, which provides a lot of information about using LEDs, along with many other components.

A Better-Looking Clock

Now that I've dealt with a night light, what else might you find on a bedroom nightstand? Very often, there's a digital clock, and it will look probably look something like the one in Figure 19-16.

Two things occur to me when I look at this clock. First, it uses transparent colored plastic. The digits are mounted behind a little red panel that could easily be acrylic. And second, the general appearance hasn't changed significantly since digital clocks were introduced in the 1970s. To me, that molded plastic has a really retro look. In which case, why not put it in an enclosure that is a bit more pleasing to the eye?

Before I get to the enclosure, I'll go into slightly more detail about the transparent red plastic. Figure 19-17 shows an additional piece of red acrylic that I stood in front of the clock. It's the same color red as the sample in the bottom-left corner of Figure 19-1. Even though the tint is quite dense, the brightness of the numerals doesn't seem to be reduced at all, but everything else is much dimmer.

The general rule is that tinted transparent plastic allows light to pass through that is the same color as itself, while it blocks other colors. This gave me the idea that if I put the clock inside a wooden enclosure, I could mount a piece of red acrylic in the front, and you'd see the numeric display of the clock quite clearly while its retro pod-shaped case would be hidden.

I decided to use poplar, which is a hardwood, but not as hard as oak or maple. It's a good compromise when you want something tougher than pine, but you'd like to avoid the labor of cutting oak with a hand saw. I figured that one-by-four board would work. Like all one-by-four wood, it is actually ¾" x 3½".

Figure 19-18 shows the concept. It's a very simple design, using just four pieces mitered at the corners like a picture frame. What could be easier?

In reality, "simple" doesn't mean "easy." Those corner joints are much more challenging than the joints in a picture frame, because they extend over the full width of the wood. This width will reveal even a slight error in cutting the 45-degree miter angles. A joint that fits at the front may not fit at the back, and if you adjust one corner so that it looks right, the other three may have problems.

The problems will be much less significant if you are using power tools, because a bench-mounted miter saw allows you to specify an angle by turning the saw and checking a pointer on a dial marked in degrees. But how can you make those cuts precisely with a hand saw?

You can't stand a piece of one-by-four wood on its edge in a typical miter box, because the sides of the box are not tall enough. If the box has a slot angled 45 degrees from horizontal, you can use this slot with the wood lying flat underneath it, but even with this guidance, sawing at a precise and consistent 45-degree angle is not easy.

A freehand cut doesn't look so good, either. If you refresh your memory by going back to Figure 9-5 on page 106, in the Swanee whistle project, you'll see how ugly that cut looked.

Figure 19-16. A typical digital clock.

Figure 19-17. Red plastic passes most of the light from red LEDs, while dimming other colors.

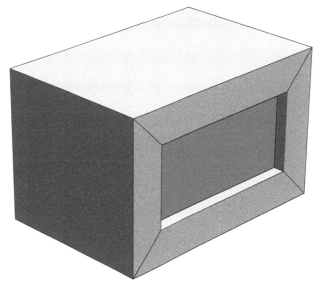

Figure 19-18. The simple design for a clock box. The sides are solid one-by-four poplar, ¾" thick.

The answer, of course, is to make a jig. In fact, I humbly present it as the Ultimate Miter Jig, which can be useful in projects other than this one. It is my final answer to the problems that I first identified in Chapter 10, when I started talking about the age-old task of connecting two pieces of wood at 90 degrees.

The Ultimate Miter Jig

Figure 19-19 shows you the jig in its complete form, to give you an idea of what I'm aiming at. Instead of requiring you to place the wood horizontally and tilt the saw at an angle, it holds the wood at an angle while the saw is vertical. I think this is much easier to use.

In the photograph, a cut has just scratched the surface of an angled one-by-four. Because the jig is in two separate pieces, they could be moved further apart to enable a miter cut in a section of one-by-six. The only limit, really, is the length of your saw.

The jig can be fabricated within a couple of hours. You need enough two-by-four pine to make two rectangles measuring 4" long, as shown in Figure 19-20, and two triangles as shown in Figure 19-21. You also need two pieces of ¾" x ¾"

Figure 19-19. Is this the ultimate miter jig?

square dowel, each 1½" long. The jig is held together with four deck screws 2½" long, and four flat-headed wood screws, size #8, 1½" long.

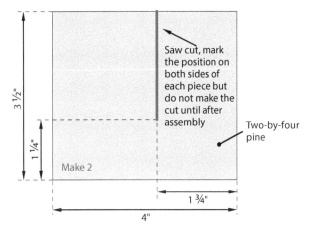

Saw cut, mark the position on both sides of each piece but do not make the cut until after assembly

Two-by-four pine

3 ½"

1 ¼"

Make 2

1 ¾"

4"

Figure 19-20. Each side of the jig is a 4" length of two-by-four.

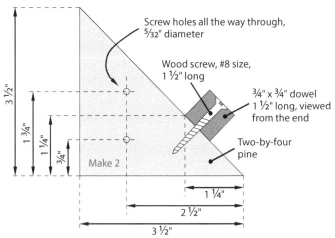

Screw holes all the way through, 5/32" diameter

Wood screw, #8 size, 1 ½" long

¾" x ¾" dowel 1 ½" long, viewed from the end

Two-by-four pine

3 ½"

1 ¾"

1 ¼"

¾"

Make 2

1 ¼"

2 ½"

3 ½"

Figure 19-21. These triangular pieces will support a work piece at 45 degrees, for mitering.

The six pieces are shown in Figure 19-22, before each set has been screwed together. Don't saw the lines on the rectangular pieces until everything has been assembled. When you countersink the holes in the triangles, remember that each triangle is a mirror image of the other.

In Figure 19-23, a piece of square dowel has been attached to one of the triangles with a #8 wood screw. I drilled a ³⁄₃₂" guide hole in the triangle before driving in the screw, to avoid splitting the pine.

The edge of the dowel, where it meets the triangle, is aligned with the vertical pencil mark on the rectangle behind it. This alignment is crucial. Make sure both pieces of two-by-four are sitting flat on your work surface, nudge them into position, and then clamp them together.

You can turn the pieces over, while they remain clamped, so that you can drive in a deck screw vertically, which is easier than doing it in a horizontal position. You need to exert some downward force so that the screwdriver bit doesn't jump out of the head of the screw.

Figure 19-24 shows a screw being inserted, but before you get that far, drill pilot holes through the screw holes, using a ³⁄₃₂" bit that sticks as far out of the chuck of the drill as possible. Then insert two deck screws to hold everything together. This will be a lot easier if you have an electric screwdriver.

After the screws are in, you can turn the pieces the right way up and make a vertical cut that will align the saw when the jig is being used. Don't ignore the thickness of the saw blade: the right-hand edge of this cut must be aligned with the edge of the square dowel, where it meets the triangle.

Figure 19-22. The parts of the jig have been cut and drilled, ready for assembly.

Figure 19-23. The triangle is ready to be attached to the rectangle of pine behind it, using a couple of deck screws.

Figure 19-24. One side of the jig being screwed together.

Figure 19-25. A saw cut, partially completed.

If the dowel shifted a little during assembly, the saw cut should go to where the dowel actually is, not where it was supposed to be. And, it is especially important that the cut must be precisely vertical, because it will determine the accuracy of the miter joints that you make with the jig. A cut in progress is shown in Figure 19-25.

Figure 19-26. Verifying the saw cuts.

Complete the saw cut all the way down to the edge of the square dowel, and then assemble the second half of the jig. After you have made a saw cut in that one too, clamp them together and resaw them as in Figure 19-26, to check that the two cuts are aligned. If they are not quite aligned, that's okay—the saw will widen them slightly, but should still maintain its correct orientation relative to the square dowels.

Does it work? I think it works better than any other options I could imagine, using hand tools. A sample cut is shown in Figure 19-27.

Figure 19-27. A sample 45-degree cut.

Note that the jig only adds a bevel to the end of a work piece that has already been cut to size. Getting back to the clock-boxing project: you have to use your miter box to prepare the four pieces of poplar with square ends, then bevel them in the jig.

Building the Clock Enclosure

The dimensions of the poplar, before and after beveling, are shown in Figure 19-28. These dimensions assume you will be using a window of red acrylic measuring 4" x 2". I am also assuming that your clock will fit into a box 3¾" wide and 1¾" high on the inside, with a usable depth of 3". If you have a larger clock, you'll need to increase the size of the enclosure and the acrylic panel.

How are you going to install the acrylic? The obvious way is to insert it in a groove around the inside of the enclosure while the enclosure is being assembled. If you are using ⅛" acrylic, you need a groove just over ⅛" wide. The dimensions I have supplied for the pieces of poplar assume that the groove will also be ⅛" deep. This is clarified in Figure 19-29.

In Figure 19-30, the edges of the groove have been penciled at ¼" from the long edge of a piece of poplar. Figure 19-31 shows two saw cuts that have been made along the pencil lines. But how can you remove the wood from between the two cuts? This would be easy with a router, but can also be done with a tenon saw, if you don't care that the bottom of the groove will be a little rough.

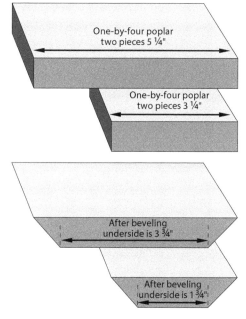

Figure 19-28. Dimensions of one-by-four poplar to make the clock box.

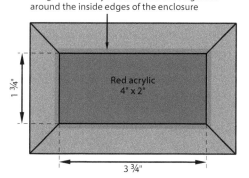

Figure 19-29. The 4" x 2" red acrylic window fits into a groove around the inside of the enclosure.

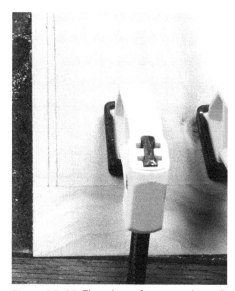

Figure 19-30. The edges of a groove that will retain the acrylic panel.

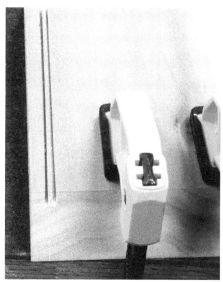

Figure 19-31. Saw cuts along the pencil lines.

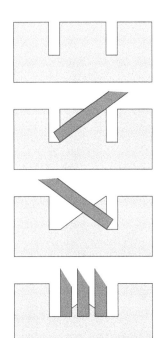

Figure 19-32. Removing wood from between two saw cuts, to make a groove.

What you can do is shown in Figure 19-32, where the top section shows the wood viewed from the end, with two saw cuts in it. Angle the saw from one cut to the other, and take out a very thin triangular slice of wood, as shown in the second section, where the gray silhouette is the end view of the saw blade. Then reverse the procedure in the opposite direction, as shown in the third section of Figure 19-32. Finally hold the saw vertically and run it through the groove from varying angles, to get rid of as much of the residue as possible, as shown in the bottom section.

The actual result is shown in Figure 19-33. The groove is slightly wider than ⅛" and slightly deeper than ⅛", to allow some room for error when gluing everything together.

Assembly is now just a matter of applying carpenter's glue to the miter joints, inserting the acrylic (which does not really need to be glued into the groove) while putting the four pieces of the enclosure together, and holding them until the glue sets. The system I used is the same as in Chapter 5, when I discussed options for clamping a frame. See Figure 19-34.

Figure 19-35 shows the finished box, with the clock inside it and the display visible. Because even a small alarm clock generates a little heat, the rear end of the enclosure can be left open to provide ventilation. If you really want to put a back on it, you should drill some ventilation holes.

Figure 19-33. The sawn groove.

Figure 19-34. Waiting for the glue to set.

Enhancements

This project suffers from one major defect: you have to remove the clock from the box to set it or stop the alarm. In the future, maybe you'll learn just enough about electronics to consider opening the case of the clock and accessing the circuit board inside. When you do that, you can run wires out from the key locations on the board to some buttons that you add to the wooden box. The clock will then become permanently installed.

Maybe that sounds a bit ambitious, but really it isn't so hard. At the risk of promoting myself too much, once again I have to mention *Make: Electronics*, which explains how to make these kinds of modifications, and is written for people who have no previous experience, just as this book was written for people who have no previous experience with tools.

Figure 19-35. The clock in a box.

Chapter 20
More Tools

Throughout this book I have tried to limit the tools, parts, and materials that were necessary for the projects. I wanted you to find out if you have a lasting interest in this area before you spent too much money on it.

Time, now, to consider your options for going beyond these limits. I'm going to make some suggestions about a workbench and storage options. Then I'll provide a list of useful additional hand tools, and I'll end by describing some fundamental power tools.

A Minimal Workbench

You can find literally hundreds of plans for **workbenches** online, so I won't go into a lot of detail, here. I'll tell you my opinions, while acknowledging that other people's opinions may differ.

I believe that the surface of a workbench should be made from ¾" plywood. OSB is another option, but it isn't as smooth. Chipboard is a possibility, but it breaks much more easily. Really, I think that plywood has the best characteristics, and doesn't have to be of the highest quality. Good-one-side is quite adequate.

If your bench isn't going to be used as a BBQ stand or as a resting place for oily car parts, I don't think you need to seal the surface. It will accumulate scuff marks, stains, and scars, but I don't think they will matter.

The bench should be as wide as space permits, but I don't think it has to be very deep. From front to back, I think 24" to 30" is about right, bearing in mind that you will be doing most of your work near the front edge. Ideally the back edge should butt up against the wall, so that items such as drill bits don't roll off and fall out of sight. If you can attach the bench to the wall for greater stability, that's an advantage.

Stability is very important. If something moves or vibrates while you're working on it, you're going to make errors. Many people use two-by-fours to make a frame around the underside of the work surface, with more two-by-fours or four-by-fours for legs. You can connect the frame with the legs with multiple 2½" deck screws driven in really tight. If you space pairs of legs at intervals of 24", the top of the bench should be strong enough for you to hammer things on it.

The height of the bench is a matter of taste. It should feel comfortable for you, depending on how tall you are. As a rough guide, I think the bench should be just below your elbow when you stand beside it with your arm bent. You want to be able to use your weight to hold things while you are working on them, but you don't want the work surface to be so low that you have to stoop to see details clearly.

If the bench cannot be anchored in the wall, you can add a back to it to increase its rigidity. A piece of ¼" plywood will do, so long as it is secured at multiple points. You can add end panels to the bench also, for stability. You want the workbench to feel absolutely solid.

I don't think it has to look beautiful. I do think it has to be strong.

Adding Work Tables

A bench is only half of the story. When you are building something, and especially when you are sawing large, flat sheets of wood, you may feel a need for a table on which to spread everything out.

When I designed my own large workshop, I decided to have two free-standing tables in the center of the area, allowing me to walk around them when building something large and heavy. Each of the tables is 48" x 48", with a 3" a gap between them, so that I can cut a sheet of plywood with the blade of a hand saw or circular saw descending into the gap. This eliminates the problem of supporting both sides of a long piece while cutting it. See Figure 20-1.

Figure 20-1. Two work tables simplify the task of making long cuts (or small cuts) across sheets of plywood or plastic.

I realize that most people may not have sufficient space for this setup, but even if you have to use smaller tables, I tend to think that two are better than one.

Tool Storage

I've seen two schools of thought on this subject. Some people like to hang tools on a wall or on pegboard, so that they're instantly available. Others like to store tools in boxes so that everything can be neatly put away in a dust-free environment.

I'm in the second category. I have at least 50 plastic storage containers, all labeled, so that I know where everything is. When I'm working on a project, I pull out just the tools that I need.

Figure 20-2. A rack for sheets of wood and plastic.

Figure 20-3. Storage for strips, tubes, dowels, and other long, thin pieces.

In theory, I like the idea of being able to grab any tool quickly, but as the number of tools increases, this scheme becomes less practical. I've seen workshops where tools were mounted on several sheets of pegboard, hung from the wall on hinges, so that you could turn them like pages in a book. To me, this isn't any quicker than opening a box, and is a less efficient use of space.

Still, it really comes down to a matter of taste. The only point I would emphasize is that whatever system you use, you'll benefit by thinking about it and planning it instead of just allowing it to happen.

Materials Storage

Are you going to buy exactly the right amount of wood for each job that you do, so that you never have any leftovers? That sounds implausible. You should always buy some extra, in case you make a mistake; and you should expect to have a few scraps left over that are too potentially useful to be thrown away.

Figure 20-2 shows a **rack** for storing sheets up to 48" x 48" square. Smaller pieces are in bins on the top, and the whole thing is on large casters, allowing easy access from both sides (and easy cleaning behind it).

You also need a way to store strips, tubes, dowels, and other long, thin pieces up to 96". Shelves are generally recommended for long pieces of lumber, but they occupy a lot of wall space and don't make it easy to sort through the wood. I prefer to stand long pieces on end, even though they may tend to bend a little.

One way to store a lot of long pieces in a small space is shown in Figure 20-3. The box contains a grid of fence wire at the top, and a matching grid below it. I drop each piece of wood, tubing, or plastic through a hole in the grid, and into a matching hole at the bottom. Of course, many other designs are possible. Maybe you can come up with one of your own.

Parts Drawers

Sooner or later, you need a better way to store fasteners than the little bags and boxes from the hardware store. A methodical system of labeling will be helpful, so that you can find what you want. I may have taken this to an extreme, with a complicated color-coded system, but on the other hand, I never have to hunt for anything anymore. See Figure 20-4.

Figure 20-4. Parts labeling.

Saw Horses

Mostly used in construction work, they are ideal if you want to take some tasks outside. Throw a cheap piece of ¾" plywood across them, and you have an impromptu work bench.

Lighting

Although it is often overlooked, bright lighting is essential for doing precise, accurate work. It is also an important safety consideration.

Additional Hand Tools

One handsaw was sufficient for all the projects in this book, with two clamps and just a handful of other tools. I don't think any other book on tools has been so frugal! What did I leave out, that would be genuinely useful?

Long Ruler

The projects that I described here have all been for small objects, for which an 18" ruler was sufficient. Sooner or later you're going to work on something bigger. A **36" ruler** is useful when cutting long shapes out of plywood, and a **48" ruler** allows you to draw a line all the way across a standard plywood sheet.

Tape Measure

Most people regard this as an essential item, but I didn't include it because it's more appropriate for home repairs, and not so easy or accurate to use in small fabrication projects. As your tool skills increase, however, you may want to fix or improve things around the house—or someone else may want you to.

Caliper

Rounding out your measurement capabilities, suppose you want to measure the diameter of a nail—or a dowel. You can hold it against a ruler, but the best you will get is an approximate idea of the number. The right tool for the job is a **caliper** (which some people refer to as **calipers**, because originally this tool used to consist of two jointed arms, like scissors).

A caliper with an electronic readout allows you to select the units of your choice, but needs a battery. A caliper with a needle against a dial may be slightly harder to read, but it doesn't need a battery. Either way, this is a really helpful measurement device, and low-end versions are available cheaply. A non-electronic example, often known as a dial caliper, is shown in Figure 20-5.

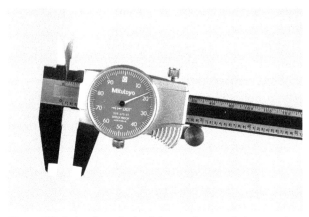

Figure 20-5. A caliper. Each of the smallest gradations on the dial is 1/1,000th of an inch.

More Clamps

I minimized the number of clamps in the projects in this book, but you probably need two large, two medium, and two small trigger clamps, plus at least one screw clamp for greater holding power. If you plan to build large objects, you need more.

More Saws

Even if you're not planning to work with metal, you need a hacksaw "just in case." At the very least, you may want to cut the end off a bolt that is too long.

I think it's also a good idea to try a Japanese pull-saw, because some people find them easier to use, and most of them don't cost a lot.

A toolbox saw of the type that I showed in the Handsaw Fact Sheet in Chapter 1 is good to have, as it can plunge right through a saw cut, unlike your tenon saw, which is restricted by the stiffening bar along the top of the blade.

Pliers

I mentioned medium-sized **slip-joint pliers** and **long-nosed electrician's pliers**, but many others exist. Those in Figure 20-6 have a double-jointed action that can apply a lot of force to small objects. The ones in Figure 20-7 are known as **lineman's pliers**, and measure about 9" long. **Locking pliers** in Figure 20-8 contain a screw adjustment that you use to set the approximate size, after which you close the pliers on the object that you want to hold, and the articulated linkage will grip it until you press a lever to release the grip. Often these are known as **Vise-Grip pliers**, as this was an early brand name.

Metal Files

Not only useful for shaping metal, they can be used with hardwood and can clean the sawn edges of plastic. If you buy a set, it usually contains a couple of flat files, a round file, and a triangular file.

Rasp

Like a metal file, but with large teeth designed for wood. The rasp is an aggressive tool that leaves a rough finish and is only useful for getting approximately the shape you want. After that, you switch to a sanding device.

Wrenches

I described some options for wrenches in the Nuts and Bolts Fact Sheet on page 180. You certainly need something that works better than an adjustable wrench. A small socket wrench set is probably the way to go, when you are beginning to buy more tools.

Also consider a set of Allen wrenches, sometimes known as hex keys. These are L-shaped six-sided rods that loosen or tighten fasteners with six-sided recesses in their heads. A full set of Allen wrenches should contain metric as well as US sizes.

Metal Shears

Like a giant pair of scissors, they are powerful enough to cut thin metal sheets. I have also used them to cut carpet and to snip plastic rods.

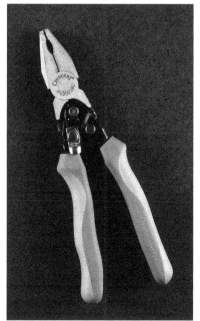

Figure 20-6. High-quality double-jointed pliers can apply a lot of force.

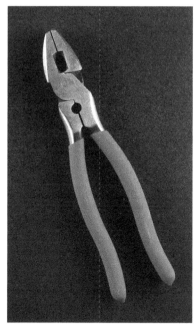

Figure 20-7. Heavy-duty lineman's pliers about 9" long.

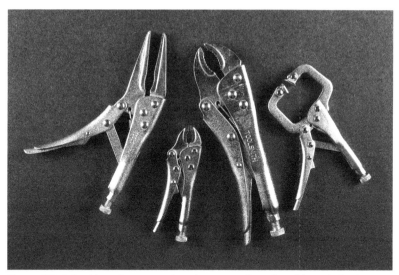

Figure 20-8. Locking pliers, sometimes known as vise-grips, although that is actually a trade name.

Bolt Cutters

If you saw the movie *Mad Max: Fury Road*, you know how useful bolt cutters can be. Seriously, I find myself using my pair about once every couple of months, especially for chopping thick wire.

Glass Cutter

Maybe you think you'll never need to cut a pane of glass, but one day, when you least expect it, you may wish you had this little tool. Its small diamond wheel scores a line into the glass. You then place the glass so that the line coincides with a firm edge underneath, and snap it along the line.

Definitely practice a couple of times before making the actual cut, and definitely wear appropriate gloves. Glass edges can be scalpel-sharp.

Bench Vise

If you acquire a workbench, you should really have a vise bolted to it. Use it as an easier alternative to a clamp when you are working with a small part or a short, thin piece of wood, plastic, or metal, and you want to grip it quickly, firmly, and reliably. A vise designed to hold metal can hold wood if you line its jaws temporarily with something slightly soft. I use an L-shaped plastic molding.

Workshop Vacuum Cleaner

Cleanup is such an essential part of shop work, you need a vacuum cleaner that lives where most of the dust comes from. In addition, your vacuum cleaner should have a hose that can attach to equipment such as power saws, to suck up dust at the source. A workshop vacuum cleaner is often referred to as a **shop-vac**, even though this is actually a brand name.

Sanding Equipment

To reduce your sanding labor, I'll mention four possibilities: a **sanding disc** for use with an electric drill, a bench-mounted **disc sander**, a handheld **orbital sander**, and a handheld **belt sander**.

Sanding Disc

This is by far the cheapest option. In its most traditional form, it consists of a flexible rubber or vinyl disc and a steel rod that goes through the center. One end of the rod can be inserted in the chuck of an electric drill, while the other end is threaded for a bolt and a cone-shaped piece of metal that holds a circle of sandpaper in place against the flexible disc. Replacement discs are cheaply available in a range grits. Figure 20-9 shows the pieces assembled, while Figure 20-10 shows them disassembled. Other sanding discs use a form of Velcro to attach the sandpaper to the flexible support, or they may use **pressure-sensitive adhesive**.

Figure 20-9. A sanding disc mounted on a drill.

Figure 20-10. The yellow disc in this photograph is made of soft vinyl.

There are two disadvantages of this tool. First, you have to switch between using the drill to make holes, and using the drill with the sanding disc. Second, it isn't ideal for sanding flat surfaces. This is because you are controlling the disc by moving the drill, and sweeping the drill to and fro in absolutely straight, even lines is difficult. You may tend to leave crescent-shaped marks in the wood that you are sanding. You need a smooth, gentle touch, and some practice.

A sanding disc is appropriate for rounding edges and (if fine-grit paper is used) removing small blemishes from flat surfaces. It is not very useful for flattening a surface that is not flat.

Bench-Mounted Disc Sander

This is different from a sanding disc in that it is a standalone piece of equipment designed to be mounted on a workbench. See Figure 20-11.

A **disc sander** allows you to bring the wood to the disc, instead of the other way around. This allows much better control. It is fitted with a table that can be oriented at 90 degrees to the disc, or in many tools can be tilted at an angle marked on a scale.

The disc is often 8" diameter or bigger. In addition, most affordable sanders of this type have a sanding belt as well as a disc. The belt is useful for sanding in line with the grain of the wood, and can flatten an area that isn't entirely flat.

A disc sander need not cost a lot of money. You probably don't really need a high-end, name-brand model. You can find one in the Harbor Freight tools catalog for the price of a family dinner at Denny's (depending how big your family is).

Figure 20-11. A bench-mounted sander allowing the options of disc or belt.

Some sanders have a bag at the back to collect dust, but in my experience, a lot of dust doesn't end up in the bag.

Do you have somewhere to put this thing? Ideally, you need a workbench where sawdust isn't a problem; but not all of us have such an ideal situation. Maybe you can store the sander in a closet, and haul it out to your back yard when you need to use it (if you have a back yard).

Breathing dust is not a good idea. A dust mask should be worn, and if you have a helper, he or she can hold the hose of a powerful vacuum cleaner near the object that you're sanding.

Figure 20-12. A handheld belt sander.

Figure 20-13. An orbital sander.

Handheld Belt Sander

If you need to smooth a piece of wood that is too big to lift and hold against a bench-mounted sander, you can clamp the wood to a workbench and run a handheld belt sander along it. This is a powerful piece of equipment that does that one task very well, but is not so versatile. An example is shown in Figure 20-12.

Orbital Sander

This moves a piece of sandpaper in little circles at very high speed. It is excellent for creating a smooth, uniform look on flat areas of wood, but not so useful for rounding edges, as it can be difficult to control. A bench-mounted disc sander is easier for that purpose.

The orbital sander shown in Figure 20-13 requires special sanding discs that use a type of Velcro to stick themselves to the sanding pad. The discs are perforated because the orbital sander functions like a little vacuum cleaner, sucking up the sawdust that it creates. Most of the dust does seem to end up in the attached bag.

Which Sander to Buy?

In order of usefulness, I put the disc sander first, then the orbital sander, then the sanding disc, and the handheld belt sander last. But the sanding disc will cost much less than the other options.

Sander Safety

A moving piece of sandpaper doesn't look very dangerous, but remember, your skin is much softer than the wood that you are shaping. If you touch a powered sander, it will immediately remove the epidermis (your surface layer of skin), and potentially can do much more. Your hand may be trapped in the gap between the table and the disc, or if you touch the spinning edge the disc, it will carve a neat furrow in your finger tip within a fraction of a second (ask me how I know).

Like all power tools, sanders must be used carefully.

Power Tools

All motor-driven tools are potentially dangerous, but some are more dangerous than others. I'll begin with the ones that seem safest to me, and work my way up.

Drill Press

A drill press is an electric drill mounted on top of a steel column, with the chuck facing downward toward a flat, rigid table. You place what you want to drill on the table, and elevate it toward the drill, usually by turning a hand crank. When your work piece is within the working stroke of the drill, you switch on the motor and pull a lever to lower the bit into the thing that you're drilling. See Figure 20-14.

The big advantage of this tool is precision. The drill bit is vertical, the work can be held at a specific angle, and a scale tells you how deeply the bit is penetrating the work piece. The speed of the drill bit can be adjusted, often by moving a drive belt between two stepped pulleys.

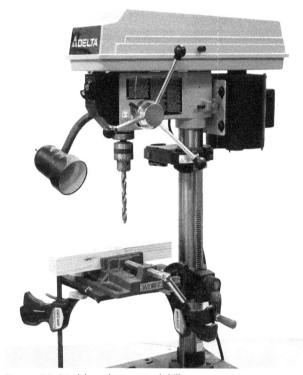

A full-size drill press stands on the floor and can be five feet tall, allowing room for large pieces of work, such as furniture. I don't think you need to spend the money on that. For less than half the price, you can get a bench-mounted drill press that will handle most normal jobs.

Figure 20-14. A bench-mounted drill press.

You will need to clamp your work while you are drilling it, and a **drill press vise** is the tool designed for this purpose. It will hold your work in correct alignment, while your drill bit descends vertically. Making the Swanee whistle in Chapter 9 would have been trivial with a drill press (but, it might have been less fun). Sometimes a vise is supplied with a drill press, but you may have to buy one separately.

Biscuit Jointer

Alignment kits are available to help you to make joints using pegs and glue, but they're still not as easy as using a biscuit jointer, which makes exact alignment unnecessary. The tool contains a horizontal circular saw blade that carves a curved slot in the edge of a piece of wood. You make a matching slot in a second piece of wood, put glue in the slots, then bring them together with an oval-shaped "biscuit" of compressed wood between them.

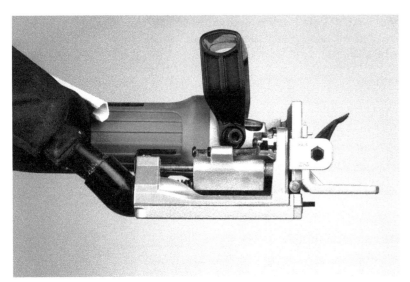

Figure 20-15. Side view of a biscuit jointer. See text for explanation.

A side view of a biscuit jointer appears in Figure 20-15. The saw blade can be seen as a black silhouette protruding about ½" at lower right. The base of the tool rests flat on the work area. Pushing the handle forward causes the blade stick out and carve a slot in the edge of a piece of wood that is also resting on the work area, immediately in front of the jointer, and under the guard at far right. In this photograph, the blade is fully extended, against the tension of a spring. When the tool is released, the blade retracts itself.

The great advantage of this system is that it allows some wiggle room. Unlike a wooden peg, which has to be precisely aligned with each hole, the biscuit can slide a little way parallel with the slot that it's in.

You can buy the necessary biscuits in quantities of 50 or 100, and they are available in a variety of sizes.

The blade of the jointer is hidden under a guard, and you should never have a chance to come into contact with it. Still, the blade is there, and would be very dangerous if you somehow got your fingers under it, or if the guard stuck in an open position.

Jigsaw

I described this handheld power tool in the Holes and Curves Fact Sheet on page 167. It's probably the easiest and most affordable of the various saws that can cut around a curve. The key to using it is to clamp your work very securely, so that the up-and-down motion of the jigsaw blade doesn't pull it loose. The exposed blade, with no guard, does make this a potentially hazardous item, although the blade is relatively short. A variety of blades can be bought, to match the type of job that you're doing.

Reciprocating Saw

This is often known as a Sawzall, as that is the brand name of a popular model. It accepts replaceable blades that can be up to 16" long, and are available for wood, metal, and other materials. When you pull the trigger, the blade oscillates in and out at high speed. See Figure 20-16.

The whole length of the blade is exposed, but if you hold the body of the tool in both hands and point the blade away from you, it should be reasonably safe. The biggest risk is that you may allow the saw to come into contact with your body or legs. This is especially relevant in battery-powered models that are "always on." If you trip and fall while using a Sawzall, the consequences could be unpleasant.

A reciprocating saw does not do precise work, but is very useful for home-repair tasks such as cutting water pipe or two-by-fours in locations that are hard to get at.

Band Saw

One type of band saw is vertically oriented, while the other type is horizontal. Both of them use a saw blade that is a continuous loop mounted on large, hidden wheels. The wheels turn continuously when the saw is turned on.

Figure 20-16.
A reciprocating saw.

The horizontal type is typically hinged, so that you open it to insert material, and then allow it to close slowly while it cuts. This type is often used for chopping pieces of long metal.

A vertical bandsaw allows you to guide your work across a rigid table and into the blade with both hands, allowing precise control. It is especially useful for cutting curves in relatively thin material,

Because your hands are resting on the table, the risk of random movements is minimized—but of course, it is not eliminated completely. A bandsaw looks relatively safe, but versions are used in butcher shops for cutting through meat and bone.

Compound Miter Saw

A powered **miter saw** is a bench-mounted tool with a circular blade that can pivot around a vertical axis, enabling you to do accurate miter cuts by aligning a pointer with a scale. It is often referred to as a **chop saw**.

In a **compound miter saw** the blade can also tilt around a horizontal axis, enabling you to make a more complex miter cut. A **sliding compound miter saw** has the extra ability to slide the blade in and out, so that you can cut wider pieces of wood. This resembles a **radial arm saw**, which is more likely to be a large, fixed piece of equipment, while a miter saw is relatively portable. An example of a compound miter saw is shown in Figure 20-17.

Figure 20-17. A compound miter saw.

Theoretically you are supposed to clamp the wood in a miter saw while making a cut, but there is always the temptation to hold the wood with your free hand. If you do this, and you don't check yourself carefully *every single time* you lower the blade, you are risking your fingers.

Figure 20-18. A typical handheld circular saw.

Figure 20-19. The chuck of the router is visible immediately above the circular hole at the bottom, and can accept a variety of bits specially designed for use with this type of tool.

Handheld Circular Saw

Also known as a Skil saw, after the manufacturer of a popular brand. This type of handheld saw is widely used on construction sites for cutting flooring and two-by-fours. Battery-powered versions are available, but the motor takes so much current, battery life is limited. A plug-in version is shown in Figure 20-18.

The blade has a spring-loaded guard that should always be in place when the saw is not actually cutting wood, but in practice, some people recklessly remove the guard—because, you know, it gets in the way. I was talking to someone recently who saw a person on a construction site run a Skil saw through the top of his leg. This is a dangerous tool when not used properly.

Router

Cutting grooves in wood and creating curved edges on lengths of wood are the special capabilities of a router. It consists of a vertical bit that points downward and can be moved in any direction through or around a flat piece of board or plywood. A wide variety of shapes of bit will create all kinds of contours. A router is shown in Figure 20-19.

This is a truly wonderful, versatile tool, but because the router will cut in any horizontal direction, there is no practical way to put a guard below the bit. A router can run at extremely high speeds, carving its way through almost anything in its path. Because this is a handheld tool, it can be dropped, or turned over, and there is the risk of pulling it out of the wood when the power has been switched off but the bit is still spinning. Because most routers are plug-in tools, there is also the risk of tripping over the power cord.

Table Saw

This is a circular saw mounted vertically in a slot in a rigid table. You can slide your work across the table and into the blade, which can be tilted if you want to make an angled edge. A fence can guide the wood at a constant distance from the blade. This type of saw is especially useful for cutting along the grain in lengthy pieces of wood. You can, for instance, convert a two-by-four into a 96" length of 1" x 3", or any other dimensions you want. Table saws are used for many other tasks in a typical workshop.

I feel that this is the most hazardous workshop tool, not just because it allows easy access to a whirling saw blade, but because the blade can grab the material and throw it at you with great force. This well-known hazard is known as **kickback**. A table saw typically has a kickback guard, which may provide some protection.

Summary

I regret that I had to take so much space warning you about power tools. I don't like to sound negative, but I'd be irresponsible if I didn't mention the risks.

Are accidents inevitable? Not at all. A man I know who teaches advanced wood-working claims that he hasn't had a student injury in 10 years. Probably that is because he watches the students, and they don't take any shortcuts because they know they are being watched.

Personally, I have used power tools for many years without any serious injury. They have enabled me to create an amazing range of objects and ornaments, which would have been impossible otherwise. I believe they can do the same for you, if you learn to use them carefully with a little expert guidance.

From the clutter and mess in a typical workshop, unique and wonderful objects emerge and find an enduring place in the world. Everything that you fabricate will have a little of your personality in it, and will be just a little different from anything made by anyone else.

I always enjoy seeing what people make, so if you come up with something interesting, please don't hesitate to send me a jpeg at platt.tools@gmail.com. In the meantime, I hope you have fun learning the skills that I have described, because it has been a pleasure for me to pass this information along.

Appendix
Buying Guide

This appendix tells you where to obtain tools and supplies, and summarizes the requirements for each project in the book.

Tools

The tools that I am recommending for the projects in this book can be divided into three groups:

Essential tools. A handsaw is the most obvious example, as almost every project requires you to cut materials with a saw. But a plastic-cutting knife is also essential if you want to score and snap acrylic plastic. The table in Figure A-1 shows which tools are essential for each of the projects. See page 250.

Optional, not essential. These tools will make your task easier, but if you're on a tight budget, you can omit them. An electric screwdriver is a good example. You can use a manual screwdriver, so long as you don't mind taking a little more time and expending a bit more effort. See Figure A-2 on page 252.

Tools for future use. Chapter 20 suggests additional tools that you can buy if you become more serious about making things in the future. Fact sheets scattered through the book also include some ideas.

Supplies

Supplies are divided into two categories: **buy-what-you-need**, shown on page 253 in Figure A-3, and **single-purchase supplies**, shown on page 254 in Figure A-4.

If you don't intend to build some of the projects, there's no point in buying supplies for them. For instance, the one-by-two wood for the bookcase in Chapter 11 is not used in any other project, so you can skip it if you're not interested in the bookcase.

Some types of supplies are sufficient for any number of projects in the book. Carpenter's glue is an example. One bottle will be enough, and you'll still have most of it left over at the end. Figure A-4 shows where each supply is used in a particular project, but you can be confident that a *single purchase* of each of the items listed will be sufficient for all the projects.

What I Hope You Have

I'm assuming that you already have letter-sized white paper, pencils, a pencil sharpener, an eraser, a fine-point black water-based roller-ball pen such as the ones made by Uniball, and a couple of other colors of any kind of pen or pencil. You should also have paper towels, and some clean rags. If you choose to use oil-based polyurethane or stain, and you make a mess, you'll need some mineral spirits or paint thinner to clean it up.

All other tools and supplies should be itemized in the tables on the following pages.

Where to Buy

I like to shop online as much as possible, but when I'm looking for tools and materials, I go to a store where I can examine the merchandise in person. This is especially true when shopping for large pieces of wood.

Wood

Many times I have sorted through a stack of two-by-fours, rejecting 80 percent of them because of their poor quality. You'll need to perform this kind of inspection of the two-by-fours and one-by-sixes that you purchase for projects in this book.

In the United States, you can shop for lumber at nationwide chains such as The Home Depot, Lowe's, Ace Hardware, or True Value, but you may find better quality if you're lucky enough to live within reach of a smaller, independent lumber yard. The only way to find out is to go and take a look.

For good-quality plywood and for the dowels that I have used extensively in this book, big-box stores and chains may have an incomplete stock of sizes, and the dowels that they sell may be made from pine. This is frustrating, because you'll get better results with dowels made from hardwood such as oak, maple, or poplar. Usually a sticker with a bar code will tell you what kind of wood you are dealing with, but pine is easily recognized by the clearly defined brown-and-cream colored stripes of its grain.

Crafts stores such as Michael's or Hobby Lobby usually stock square and round dowels, and will also have small pieces of high-quality plywood, such as the ⅛" birch required for the parquetry project in Chapter 7. The prices in crafts stores tend to be relatively high, though.

Adding it up, your best option for relatively small pieces of quality wood may be to shop online. A wide range of hardwood dowels and hardwood-veneered plywood can be found from suppliers selling through Amazon. You can also buy from carpentry supply sites such as www.woodcraft.com. A search engine will turn up more options.

Other Materials

For plastic supplies, I like to use TAP Plastics at www.tapplastics.com or United States Plastic Corp at www.usplastic.com. TAP has useful videos on its site, and both companies employ knowledgeable telephone help. Don't hesitate to pick up the phone and ask questions. But if you just want pieces of ABS or acrylic that are 12" x 12" or 12" x 24", eBay may be less expensive.

In fact, eBay is often my starting point for almost any crafts-related browsing.

Fasteners

All the nails, bolts, and screws for projects in this book should be stocked in any hardware store, except for the very small #2 size screws that I specified in a couple of projects, and the partially-threaded bolts that I suggested for the pantograph in Chapter 14.

One source can always be guaranteed to have these and any other fasteners that you want: www.mcmaster.com, the web site for McMaster-Carr, which probably stocks a wider range of hardware than any other supplier in the world. They aren't the cheapest, but I believe they are the best, not only in terms of their inventory but in their speed of delivery, their customer service, and the general information on their web site.

Tools

For low prices, Harbor Freight Tools is widely recognized as a bargain-basement source. They have more than 700 retail outlets, or you can order from www.harborfreight.com. Buying multiple tools and supplies from a single source can save time, but inevitably, cheap products are not always the best. Regular customers of Harbor Freight claim that it's a great place for some things, but not others. I can't advise you on that, but if you search online for "harbor freight reviews," you'll find a lot of useful information.

Sears is still a viable retail source for moderately priced tools (at the time of writing), while Northern Tool sells name brands affordably at www.northerntool.com.

Sorry, No Kits!

If you were hoping that I could offer you a kit of tools and supplies ready-made for this book, I have to disappoint you. Stocking and shipping a collection of heavy items such as clamps and saws, and 8-foot lengths of two-by-fours, was obviously not practical.

When I started writing the book, I knew that you would have to do your own shopping, so I tried to minimize the requirements. I am hoping that you should be able to get what you need without too much trouble. As a general rule, I suggest you search Amazon and eBay first, to get an idea of what's available and how much it costs. Then go shopping for the tools and supplies that you prefer to examine in person, and buy the remaining things online.

Essential Tools

#	Tool	1	2	3	4	5	6	7	8	9	10	11	12	13	14	15	16	17	18	19
1	Tenon saw with hardened teeth	●	●	●	●	●	●	●	●	●	●	●		●	●	●	●	●	●	●
2	Coping saw with 1 package of spare blades													●						●
3	Miter box	●	●	●		●	●			●	●	●		●	●	●	●	●	●	●
4	Trigger clamps: 2	●	●	●	●	●	●	●	●	●	●	●		●	●	●	●	●	●	●
5	Ruler, 18", stainless-steel, cork back	●	●	●	●	●	●	●	●	●	●	●	●	●	●	●	●	●	●	●
6	Speed square, 7"	●	●	●	●	●	●	●	●	●	●	●		●	●	●	●	●	●	●
7	Sanding block, rubber	●	●	●	●	●	●	●	●	●	●	●	●	●	●	●	●	●	●	●
8	Work gloves	●	●	●	●	●	●	●	●	●	●	●		●	●	●	●	●	●	●
9	Awl or pick (also known as scratch awl)			●	●		●	●	●	●	●	●	●	●	●	●	●	●	●	●
10	Permanent marker, fine point			●															●	
11	Hammer				●	●			●					●						
12	Pliers, long-nose or medium-size slip-joint				●	●			●			●	●	●	●					
13	Utility knife					●				●	●			●		●	●	●	●	
14	Knife for snapping acrylic plastic																			●
15	Electric drill and set of drill bits								●	●		●	●	●	●	●		●	●	●
16	Countersink, $1/2$" uniflute								●	●		●	●	●	●	●		●	●	●
17	Screwdriver, manual, Phillips size 2									●	●	●	●	●						●
18	Screwdriver, manual, flat blade, medium														●					
19	Screwdrivers, miniature set																●	●		
20	Sewing needle (carpet needle preferred)											●								
21	Metal files, set of 1 each flat, round, half-round													●		●				●
22	Wrench, adjustable, length 4" to 6"														●	●				
23	Heatgun, small type, 350W approximately														●				●	●
24	Solvent applicator, squeeze-bottle with needle															●			●	●
25	Goggles for protection against liquids															●			●	●

Figure A-1. Essential Tools.

Notes Regarding Essential Tools

These notes are keyed to the line numbers in the table in Figure A-1.

1. Tenon saws, also known as back saws or miter saws, often don't have hardened teeth, but I consider this feature essential to minimize your effort. Stanley FatMax 17-202 is my preferred choice. Other options are Pro-Grade 31968, the Irwin Plus 955, or the Silverline Tri-Cut model 456935, although this last one is shorter than I would prefer.

2. I doubt you gain any advantage from spending more than the minimum on this type of saw.

3. You will need a miter box that has a 67.5-degree slot in addition to the usual 45-degree and 90-degree slots. The 67.5 slot may be referred to as a 22.5 slot, as explained on page 65.

4. If you buy clamps that are significantly longer or shorter than the recommended 12" size, they may not fit some of the projects. The 12" dimension refers to the maximum distance between the jaws of the clamp.

5. The ruler should have square ends with no margin at each end. It must be graduated in inches in ½", ¼", ⅛", and 1⁄16" fractions, with a separate scale for millimeters. It must be 18" long (a 12" ruler will be too short).

11. Buy the weight of hammer that feels comfortable to you.

14. This may be listed as an "acrylic knife." It will be cheapest on eBay or at an online store selling plastics.

15. The set of drill bits listed on page 93 is optimal for this book. Several manufacturers sell this range of sizes as 14-bit sets. Because none of my projects entails drilling materials harder than wood, low-cost bits are acceptable.

16. The countersink must have only one flute and must be ½" diameter. Very low-cost options are available on eBay.

19. The miniature screwdrivers must include a Phillips size 1. The Stanley 6-piece set, product number 66-052, is my choice.

21. The files will only be used on wood and plastic in this book, so you don't need to pay a lot for quality. A set must include one flat file, one round file, and one half-round file. They should be at least 8" long, not including the handle. Don't buy miniature files or needle files. Sears product 22015HNN is an example, but cheaper sets are on eBay.

23. You don't need a large heat gun, so search for a 350W model. The HG-300D from NTE Electronics is my choice. Cheaper options are available, but I haven't tried them.

24. Buy the solvent applicator from the same source where you buy the plastic solvent cement (see item 26 in Figure A-4).

25. Goggles provide more side protection than safety glasses. This is important. Check on Amazon or eBay.

Tools That Are Optional and Not Essential		Chapters																			
		1	2	3	4	5	6	7	8	9	10	11	12	13	14	15	16	17	18	19
1	Dust mask	●	●	●	●	●	●	●	●	●	●	●		●	●	●				
2	Safety glasses	●	●	●	●	●	●	●	●	●	●	●	●	●	●			●		
3	Heat-resistant gloves														●				●	●
4	Utility saw with hardened teeth			●	●		●	●			●			●						
5	Electric screwdriver and screwdriver bits										●	●	●	●	●					
6	Deburring tool															●	●	●	●	●
7	Protractor					●		●												
8	Plastic template for drawing holes			●																
9	Ratchet strap					●														
10	Tape measure												●							
11	Stud finder												●							
12	Level												●							

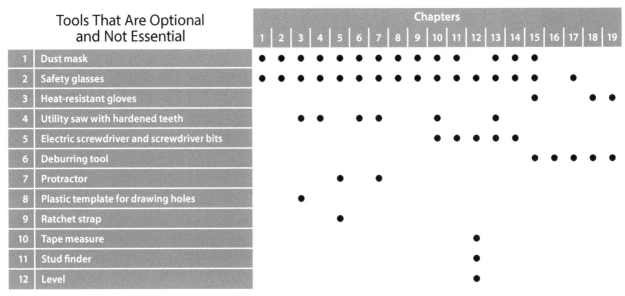

Figure A-2. Tools That Are Optional and Not Essential.

Notes Regarding Optional Tools

These notes are keyed to the line numbers in the table in Figure A-2.

1. For protection against sawdust, a low-cost dust mask is adequate.

2. Because you will not be doing metal work, I think low-cost safety glasses are probably adequate.

3. Oven mitts or the type of pure-cotton gloves used in hair salons will work.

4. Stanley model 20-221 is a good option. Just make sure that any saw you buy has hardened teeth.

6. Several types of deburring tool are available. The type you want should look like the photograph in Figure 15-1 on page 184. You won't be using it on metal, so it does not have to be of high quality.

7, 8. These plastic drawing aids are available from stationery suppliers.

Notes Regarding Buy-What-You-Need Supplies

These notes are keyed to the line numbers in the table in Figure A-3. Items 17 and 18 in parentheses are optional.

1, 2, 3, 4. To obtain clear (knot-free) two-by-four and one-by-six pine, buy common pine and saw pieces out of it between the knots. If you buy three eight-footers of two-by-four and three eight-footers of one-by-six, you should have enough for all the projects in the book.

5. If you can't buy 40", buy two pieces measuring 18" or 24" each.

6, 7, 8, 9, 10. Go to your big-box store first, where you can select dowels that aren't warped. Try to avoid pine, especially for the thinner dowels. Look for dowels made of maple, poplar, or oak.

#	Buy-What-You-Need Supplies	1	2	3	4	5	6	7	8	9	10	11	12	13	14	15	16	17	18	19	Total
1	Two-by-four pine, any condition			12"	12"		12"	18"		18"	18"		9"			12"	12"	12"	12"	18"	165"
2	Two-by-four pine, clear, not warped			9"					7"	10"					12"						38"
3	One-by-six pine, a few knots, not warped											96"		18"		12"	12"	12"	12"		162"
4	One-by-six pine, clear, not warped				12"	12"			6"							12"					30"
5	One-by-two oak, maple, or poplar										40"										40"
6	Dowel, square, ¼" x ¼"				36"																36"
7	Dowel, hardwood, square, ½" x ½"									18"											18"
8	Dowel, square, ¾" x ¾"	36"	24"		3"	60"	30"			3"			6"	12"		3"				3"	180"
9	Dowel, hardwood, round, ⅜"									12"			12"								24"
10	Dowel, hardwood, round, ¾"									18"			6"								24"
11	"Hobby board", maple, ¼" x 3½"														36"						36"
12	Plywood, birch, ⅛" thick, 12" wide							12"													12"
13	Plywood, pine or oak, ¼" thick, 12" wide						18"			18"											36"
14	Plastic, white ABS, ⅛" thick, 12" wide															24"	12"	12"	6"	18"	72"
15	Polycarbonate, ¹⁄₁₆" or ³⁄₃₂", 12" wide																	24"	4"		28"
16	Acrylic, ⅛" thick, 4" (4 colors, see notes)																		16"		16"
17	(PVC rod, ³⁄₁₆" round, 4" lengths)																		(90)		360"
18	(Optional: PVC plumbing ½" caps)																		(60)		60

Figure A-3. Buy-What-You-Need Supplies.

If necessary, try a crafts store (which will be more expensive) or shop online. Amazon sells many types of dowels.

11. Same advice as for the dowels, above.

12. You need plywood with a birch veneer (the outermost layer), because the project requires the pale color to contrast with darker tints created with stains. Try a crafts store, or buy online.

14. I specified a 12" width for ABS, because this is a common size. Check plastic suppliers, or try eBay. You want the type of ABS which is textured on one side.

15. Clear polycarbonate, also sold as Lexan, is at your local hardware store, sold as window glazing. It may be cheaper online.

16. You need a minimum of 4" x 4" of dark transparent red acrylic, ⅛" thick, and the same size of dark transparent blue, and of transparent magenta (also sold as pink). In addition, you need a piece of any other ⅛" acrylic for scoring-and-snapping practice. Currently, TAP Plastics sells 4" x 4" samples that are ideal.

17. A source of 3D printing supplies will sell you a bundle of PVC rods, ³⁄₁₆" diameter, for a modest price.

18. PVC plumbing caps for ½" pipe are much less expensive if you buy in bulk. Search online.

Single-Purchase Supplies

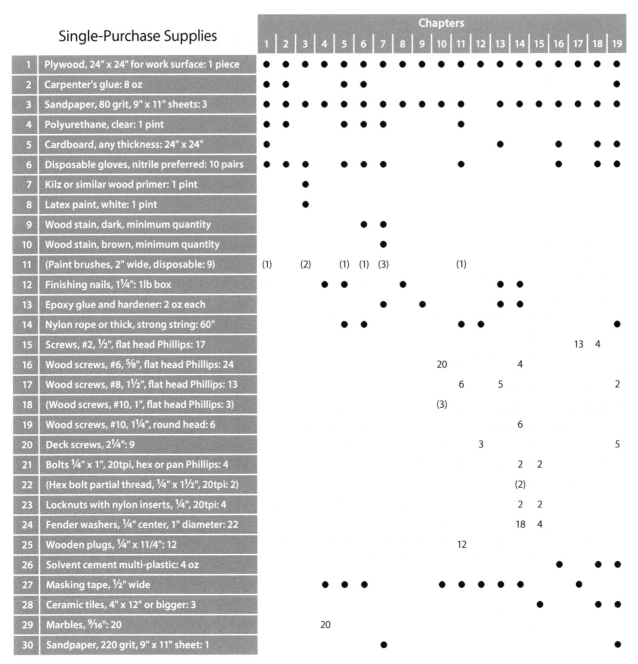

#	Single-Purchase Supplies	1	2	3	4	5	6	7	8	9	10	11	12	13	14	15	16	17	18	19
1	Plywood, 24" x 24" for work surface: 1 piece	●	●	●	●	●	●	●	●	●	●	●	●	●	●	●	●	●	●	●
2	Carpenter's glue: 8 oz	●	●		●															●
3	Sandpaper, 80 grit, 9" x 11" sheets: 3	●	●	●	●	●	●	●	●	●		●		●	●	●	●	●	●	●
4	Polyurethane, clear: 1 pint	●	●		●	●	●					●								
5	Cardboard, any thickness: 24" x 24"	●												●			●		●	●
6	Disposable gloves, nitrile preferred: 10 pairs	●	●	●		●	●	●				●					●		●	●
7	Kilz or similar wood primer: 1 pint			●																
8	Latex paint, white: 1 pint			●																
9	Wood stain, dark, minimum quantity							●	●											
10	Wood stain, brown, minimum quantity								●											
11	(Paint brushes, 2" wide, disposable: 9)	(1)	(2)			(1)	(1)	(3)				(1)								
12	Finishing nails, 1¼": 1lb box					●	●			●				●	●					
13	Epoxy glue and hardener: 2 oz each							●		●				●	●					
14	Nylon rope or thick, strong string: 60"						●	●				●	●							●
15	Screws, #2, ½", flat head Phillips: 17															13	4			
16	Wood screws, #6, ⅝", flat head Phillips: 24										20				4					
17	Wood screws, #8, 1½", flat head Phillips: 13											6		5						2
18	(Wood screws, #10, 1", flat head Phillips: 3)										(3)									
19	Wood screws, #10, 1¼", round head: 6													6						
20	Deck screws, 2¼": 9												3							5
21	Bolts ¼" x 1", 20tpi, hex or pan Phillips: 4													2	2					
22	(Hex bolt partial thread, ¼" x 1½", 20tpi: 2)													(2)						
23	Locknuts with nylon inserts, ¼", 20tpi: 4													2	2					
24	Fender washers, ¼" center, 1" diameter: 22													18	4					
25	Wooden plugs, ¼" x 11/4": 12											12								
26	Solvent cement multi-plastic: 4 oz																●		●	●
27	Masking tape, ½" wide					●	●	●				●	●	●	●			●		
28	Ceramic tiles, 4" x 12" or bigger: 3															●			●	●
29	Marbles, 9⁄16": 20				20															
30	Sandpaper, 220 grit, 9" x 11" sheet: 1							●												●

Figure A-4. Single-Purchase Supplies.

Notes Regarding Single-Purchase Supplies

These notes are keyed to the line numbers in the table in Figure A-4. Items 11 and 22 in parentheses are optional.

1. Your big-box hardware store should have precut pieces this size.

6. The cheapest disposable gloves seem to be those that are sold for food preparation. They're good for when you're painting, but may dissolve in plastic solvent. Nitrile gloves are a safer bet. Latex gloves can eventually cause a latex allergy, and should be avoided.

11. If you choose to apply paint and polyurethane using rags or paper towels, brushes will not be necessary.

13. You should find that four ounces of epoxy (and four ounces of hardener) cost very little more than two ounces. Epoxy glue has an indefinite shelf life.

15, 16, 17, 18, 19, 20, 21, 22, 23, 24, 25. Quantities specified for these items are exact. No allowance has been made in case you lose a screw or washer. I suggest that you should add some extras, just in case.

15. Your easiest source for these little screws is www.mcmaster.com.

16, 17, 18, 19, 21, 23, 24. You can probably find small quantities of these items in little bags on hooks at your local hardware store, but they are much cheaper per-piece in quantities of 50 or 100, usually sold in boxes.

20. Deck screws are sold by the box, like nails.

22. For partially-threaded bolts, I would go to www.mcmaster.com.

25. Search for "grooved dowels" on eBay or Amazon.

26. You need solvent cement that will work on ABS, polycarbonate, and acrylic, and it should be water-thin in consistency. SciGrip 3 works on all three plastics. It contains dichloromethane (also known as methylene chloride), and smaller amounts of trichloroethylene and methyl methacrylate monomer. Many suppliers sell it currently online. If it becomes unavailable, call a plastics vendor and ask their recommendation. A can of four fluid ounces will be sufficient. Wherever you buy it, you should also buy the squeeze-bottle dispenser with syringe needle. Don't try to use this solvent in an unapproved dispenser, because it may dissolve the dispenser.

29. The size of the marbles must match the spacing of the nails in Chapter 4. You should be able to find marbles $\frac{9}{16}$" diameter on eBay at a moderate price. If you buy marbles of a different size, you will have to change the nail spacing in this project.

Index